Human–Computer Interaction Series

More information about this series at http://www.springer.com/series/6033

Adrian David Cheok · Kasun Karunanayaka

Virtual Taste and Smell Technologies for Multisensory Internet and Virtual Reality

Springer

Adrian David Cheok
Imagineering Institute
Iskandar Puteri, Johor
Malaysia

and

City, University of London
London
UK

Kasun Karunanayaka
Imagineering Institute
Iskandar Puteri, Johor
Malaysia

and

City, University of London
London
UK

ISSN 1571-5035 ISSN 2524-4477 (electronic)
Human–Computer Interaction Series
ISBN 978-3-319-73863-5 ISBN 978-3-319-73864-2 (eBook)
https://doi.org/10.1007/978-3-319-73864-2

Library of Congress Control Number: 2018938655

Printed on acid-free paper

This Springer imprint is published by the registered company Springer International Publishing AG part of Springer Nature
The registered company address is: Gewerbestrasse 11, 6330 Cham, Switzerland

This book is dedicated to my wonderful daughter Kotoko. I love you infinity percent.

Adrian David Cheok

To my beautiful family (Rusandu & Kishori) and loving parents (Chandrapala & Chandrawathie).

Kasun Karunanayaka

Acknowledgements

This book represents the work done together with my research staff in the Imagineering Institute, Malaysia. Over the years, I had great pleasure working with several very hard working, talented and creative researchers in Malaysia. The research described in this book very often required many high-pressured late night, weekend and all night works to meet deadlines such as conference paper submissions, or preparing for international demonstrations, and I really appreciate such great dedication and hard work. Not all researchers can be so dedicated, and many give up, so I am extremely happy to see those researchers who passionately believe that we should aim for quantum step innovations and blue sky inventions, rather than do incremental research. This way we can help change society for the better in the future, which should be the ultimate aim of researchers.

I am deeply grateful for the great work of each researcher who helped carry out research, and also for the help writing the chapters. Apart from my co-author Kasun Karunanayaka, each researcher, who contributed to the project, is listed in Table 1.

Without the support of sponsors, the research work described in this book would not have been possible to carry out. Thus, I would like to thank Khazanah Nasional Berhad, Malaysia for supporting and being major funders of our work throughout the years.

Iskandar Puteri, Malaysia
London, UK
December 2017

Professor Adrian David Cheok
Imagineering Institute and
City, University of London

Table 1 List of co-authors

Author name	Contributed chapters	Nationality	Photo
Surina Hariri	Chaps. 6 and 7	Malaysian	
Nurafiqah Johari	Chaps. 4, 5 and 7	Malaysian	
Hanis Camelia Muhar	Chaps. 4, 6 and 7	Malaysian	
Halimatuss Saadiah Rosdin	Chaps. 6 and 7	Malaysian	
Hamizah Shahroom	Chaps. 6 and 7	Malaysian	

(continued)

Table 1 (continued)

Author name	Contributed chapters	Nationality	Photo
Sharon Kalu Joseph Ufere	Chaps. 1–3, 6 and 7	Nigerian	

Contents

Acronyms

AC	Alternating Current
ADP	Adenosine Di-phosphate
AMP	Adenosine Monophosphate
AOS	Accessory Olfactory System
ATP	Adenosine Triphosphate
ATS	Analysis of Variance-Type Statistic
cAMP	Cyclic Adenosine Monophosphate
CO_2	Carbon Dioxide
DC	Direct Current
DNA	Deoxyribonucleic Acid
DRD2	Dopamine D2 Receptor
ENaC	Epithelial Sodium Channel
FTDI	Future Technology Devices International
GERD	Gastroesophageal Reflux
GLUT2	Glucose Transporter 2
GMP	Guanosine Monophosphate
GNAT3	Gustducin Alpha-3
GND	Ground
HCI	Human Computer Interaction
HCl	Hydrochloric Acid
IC	Integrated Circuit
IMP	Inosine Monophosphate
KCl	Potassium Chloride
LED	Light-Emitting Diode
M	Mean
mGluR1	Metabotropic Glutamate Receptors 1
mGluR4	Metabotropic Glutamate Receptor 4
MSG	Monosodium Glutamate
MOB	Mammalian Olfactory Bulb
NaCl	Sodium Chloride

NO	Nitric Oxide
NPN	Negative Positive Negative
OS	Olfactory System
OSN	Olfactory Sensory Neurons
ORN	Olfactory Receptor Neurons
PC	Personal Computer
PCB	Printed Circuit Board
PG	Periglomerular Cells
PID	Proportional Integral Derivative
PKD	Polycystic Kidney Disease
PROP	Propylthiouracil
PTC	Phenylthiocarbamide
PWM	Pulse Width Modulation
QTL	Quantitative Trait Loci
SD	Standard Deviation
SMS	Short Message Service
SNPs	Single Nucleotide Polymorphisms
SOA	Sucrose Octaacetate
TR	Taste Receptors
TRC	Taste Receptor Cells
TRP	Transient Receptor Potential
TT	Thermal Tasters
TRPM5	Transient Receptor Potential Cation Channel Subfamily M Member 5
T&T	Toyoda and Takagi's perfumist's strip method
USB	Universal Serial Bus
3D	3-Dimensional

Prologue: The Mouth's Glasses

It is interesting to see how glasses are part of our reality. We internalise them as part of our face once we wear them. I believe that when glasses were first invented in the fourteenth century, it did not cease to cause surprise to the human body. It was an addition that served to mitigate our defects or increase our capabilities. I am sure that many were opposed to it and looked at those who wore them with surprise. Human beings tend to notice differences before similarities. They tend to differentiate the new as foreign. This mechanism is no more than part of a survival system; to approach people we resemble both socially and emotionally in order to preserve our genes.

From the moment humanity started to invent, it created tools to improve its life. Among these tools, they emphasised on those that enhanced our sensory capacities. Is it not the invention of better eating utensils, that became the tools that increase our sense of taste? Over the years and with the eruption of newly formed ICT tools, they became, often invisible, digital devices. Augmented reality is a fact, and we forget that it has been with us for a long time, transforming our lives by leaps and bounds. Those who are opposed to it are opposed to a vector of our historical development. Remember, nevertheless, that all great advances always have great detractors. And just like a person with glasses, we will end up assuming it as something normal.

There exist theories about the transhumanism where we are evolving as a species: a technological evolution that will allow us to acquire or improve our genetic abilities, thanks to artificial complements. At the philosophical level, this can be very interesting since it forces us to consider the limits that may stop human beings. However, without becoming dystopian, we may already be at the hinge of this paradigmal shift. Do we not already use our smartphones as an extension of our memory and our cognitive skills? Every day, we automate processes, thanks to the same device. We get up due to an alarm clock; it reminds us of an automated agenda that tells us what we have to do. If we arrive late to work, we will look for an alternative route, or it will indicate us what we should or should not read according to our tastes. In a few years, the smartphone will probably be already implanted in some way under our skin. In this world of the near future, we must understand that these increases will improve our capabilities and make us live longer, if not better.

Glasses for the mouth are only a concept. With age, human beings are losing their sense of taste and smell often causing less hedonic pleasure. Amongst the elderly, this causes many people to begin to eat less than one needs with the consequent loss of nutrients that derives in diseases. It can also be worrisome when one sees reputable chefs complain that a product does not taste the same as when they first started cooking. However, they have not realised that perhaps what may have changed is not the world, but oneself. To have the possibility of not only being able to alleviate our gustatory decadence but also to increase our capacities is something that is discussed in this book. We have seen the horizon, and I am sure that this will soon go further.

Adrian is not only a brilliant engineer, but he is a dreamer. He likes to reach out where no one has arrived yet, and he moves along the path of the dream, devising and solving problems that have not yet been raised. The more I know him, the more I am surprised, and I think that more projects like the Imagineering Institute should exist in order to reexamine humanity in post-material times.

(Credits of the picture: Alex Iturralde)

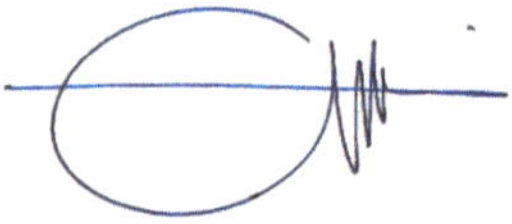

Errenteria, Gipuzkoa, Spain
December 2017

Chef Andoni Luis Aduriz
Mugaritz

Biography of Chef Andoni Luis Aduriz

What does luxury mean? Do we eat to satisfy our hunger? How are cultural identities built? Is flavour the only important thing in gastronomy? Which is the real distance between global and local? Andoni Luis Aduriz (San Sebastin, 1971) is undoubtedly one of the most influential chefs of our times. He has lead since 1998 Mugaritz, a project that he defines as a creative ecosystem that allows the freedom to create without chains and to constantly ask new questions. Throughout his career, he has prioritised both culinary evolution and an interdisciplinary approach. This has allowed him to cross the established borders, and to become a rebel in the kitchen. His lectures in places such as Harvard University or the Massachusetts Institute of Technology (MIT), his articles in El Pas newspaper and his membership to the Basque Culinary Center Foundations patronage or to the Tufts Nutrition Council from TUFTS University (a group of international leaders from diverse backgrounds who share a passion for nutrition and health), and his books are powerful tools to share all the knowledge that Mugaritz has acquired about creativity in organisations, health, perceptions or the gastronomy of the future. His versatility and creativity always raging the predefined limits open windows to many new worlds. This pioneering attitude gives rise to theatre performances with La Fura dels Baus and to the promotion of diverse documentaries such as Mugaritz BSO and OFF-ROAD. Aduriz aims to seduce us with a multisensory experience. During his professional life, he has earned prizes such as the Spanish National Gastronomy Prize, the St. Pellegrino Chefs Choice Award from St. Pellegrino and the Eckart Witzigmann Prize.

Chapter 1
Introduction

Our goal is to transform Internet and virtual reality communication into a multisensory experience by the digitization of the smell and taste senses. This will allow people to create, communicate, and regenerate multisensory information. This book was written in an attempt to present the state of the art in research in multisensory communication. Both human sense of smell and taste performs an important role in enhancing one's everyday life experiences through emotions and memory. The memory of smell and taste lasts longer than memory attained verbally. Studies have shown that smell and taste senses are directly associated with one's mood, stress, retention, and memory functions. The use of the Internet as a communication medium grew rapidly across the past two decades. To pursue the next stage of the Internet, humans should not only communicate using visual, audio, and tactile stimuli but also with smell, and taste. Virtual stimulation of smell and taste is considered a useful step in expanding the technology related to digital multisensory communication, and virtual reality.

New improvements in research will soon made it possible to connect real world things, people, plants, animals, robots, etc. together and generate an ever-increasing flow of digital data. Digitizing the taste and smell senses will lead to a technological revolution, mainly by enabling people to sense virtual smell and taste sensations, communicate them digitally over the Internet, and effectively regenerate them in remote locations. Digital smell and taste interfaces will open up a multitude of new horizons and opportunities for research in the future, including in areas of human computer interfaces, entertainment systems, medical and wellness. Digital controllability of the sensation of taste and smell provides a useful platform for engineers, food designers, and media artists towards developing multisensory interactions remotely, including the generation of new virtual tastes and smells.

This book is divided into seven chapters. This Chapter begins with the introduction, highlighting the concepts, the aims, and brief summary of each chapter.

A. D. Cheok and K. Karunanayaka, *Virtual Taste and Smell Technologies for Multisensory Internet and Virtual Reality*, Human-Computer Interaction Series,
https://doi.org/10.1007/978-3-319-73864-2_1

Chapter 2 details the basic science of taste. Taste is the key sensory modality through which we evaluate whether a potential food is good or harmful. All the basic taste qualities (sweet, bitter, sour, salty, and umami) can be elicited from all the regions of the tongue that contains the taste buds. The chapter begins with the general description of the gustatory system and its make up. It further describes the taste qualities, its receptors, genetic make up and how humans taste i.e. from the tongue to the brain. The chapter also discusses how one can adapt to a particular taste, different types of taste disorders, and how it affects ones general living. The chapter ends by describing some taste technologies that uses chemical approach and few new technologies currently using electrical approach.

Chapter 3 explores the science of olfaction. The sense of smell in many ways remains the least understood of the sensory modalities. For both animals and humans, it is one of the important means by which our environment communicates with us. This chapter discusses the anatomy of the olfactory system, dealing with the olfactory epithelium, olfactory bulb and describing how human olfactory structures differ from those of other mammals. It describes how humans perceive smell including the characteristics of olfaction. For example describing the sense of smell as a "hidden, normative, near, emotional sense" which also has a special memory. The chapter also discusses olfactory adaptation, causes of olfactory dysfunction and detrimental effects of olfactory loss. And lastly, highlights some olfactory devices using chemical approach.

Chapter 4 discusses our studies on electric taste. In the multisensory communications, sensing and actuating taste digitally is an extremely important requirement, hence the need to digitize the five basic tastes sweet, sour, bitter, salty and umami (savoury). The early sweet taste interfaces developed in the field of Human-Computer Interaction have primarily used chemicals to generate taste sensations. Electric taste interface is an electrical tongue actuator stimulation device, which the user places in the mouth and it is controlled by the computer where it can effectively generate electric taste sensations. Our previous user studies have suggested by changing the current and frequency we can produce sour, salty and bitter sensations. Once this goal is achieved, people will be able to experience taste digitally across the Internet as we experience text, audio and visuals online.

"Thermal Taste Machine" that can produce sweet sensations on the tongue is discussed in Chap. 5. This is achieved by changing the temperature on the surface of the tongue (from 25 to 40 °C) within a short period of time by placing a silver plate on the tongue and changing its temperature using a computer controlled Peltier module. This device is useful in two ways. Firstly, it can be used as a way to enhance the sweetness of the things we eat. Secondly, it also produces sweet sensations without the aid of any chemical (like sucrose) for individuals who are sensitive to thermal sweet taste. This is done by activating a special taste channel called Transient receptor potential cation channel subfamily M member 5 (TRPM 5) which produce or modify sweet taste sensations.

Chapter 6 describes the technology of digitizing smell. Until this date, almost all the smell generation methods used in both academia and industry are based on chemicals (odor molecules). These methods have limitations such as being expensive in long term use, complex, need routine maintenance, requiring refilling, less controllability, and lack of uniform distribution of odor molecules. Furthermore, these chemical based smells cannot be transmitted over the digital networks and regenerate in a remote place as we do for visual and auditory data. Therefore, discovering a method that produce smell sensations without using chemical odorants, is becoming a necessity. In this chapter, we discussed how we can bypass the chemical molecular layer of the human olfactory system and produce electric smell sensations. First we apply a minimal electrical stimulation on the smell receptors and study whether this approach can produce or modify smell sensations. This approach is similar to electrogustometry, which produces different taste perceptions by stimulating the taste receptors in the tongue using electric pulses.

And the final chapter, Chap. 7 discusses the benefits of the technologies introduced in different fields, especially in multisensory digital communication, virtual reality, entertainment, online shopping and in medical uses. It highlights some future applications and limitations as well. Throughout this book a number of possible technologies such as digital taste, thermal taste and digital smell which can bring the change from chemical based smell and taste to digital will be discussed. The technology of sharing smell and taste via the Internet is a challenging experience. As described in Chaps. 4, 5, and 6, we have produced our stimulation protocols built from literature and user studies. Furthermore, this book serves as a reference for researchers and students and generally for individual interested in digital multisensory studies because it combines the concepts, descriptions, experimental methods for digitizing the sense of smell and taste. We hope these technologies will further improve and will be used in the future of virtual reality, and Internet communication.

Chapter 2
Science of Taste

2.1 Introduction

This chapter is a general review on the science of taste. Taste, is among the five long established senses classified under the sensory system. It is the key sensory sense through which we assess if a particular food is good or harmful. Taste is the sensation produced typically in the tongue when a substance reacts chemically with taste receptor cells situated on taste buds in the mouth. The oral cavity is the portal through which everything we eat passes, so it is a common scene for evaluating and sensing what ought or not be digested and absorbed into the blood [1]. Thus, taste, together with smell and trigeminal nerve stimulation regulate flavors of food or other substances. If the potential food is good and acceptable, other taste-cued reflexes such as the exocrine and endocrine secretions which are triggered by taste get involved to expedite assimilation of the available nutrients [2].

Brain stem reflexes are evident in humans prenatally, in that, acceptance of sweet taste indicates calories, while the refusal of strong bitter taste indicates toxins [2, 3]. Food precedence are based on these reflexes, which could be altered by experience, however they are never eradicated. Irrespective of whether taste-guided behaviors are reflexive or are part of a more advanced developed appetite, the taste system must identify what is presented in the oral cavity and allow perception and recognition of chemical components there. These procedures eventually result to the acknowledgment of food as well known, or when new as harmless. Feeding is often always disturbed, when taste is extremely irritated [4].

The main outward sensory system essential for life is arguably the sense of taste. Without sight, smell or hearing, people may still live a normal life in our societies. With visual feedback, even some somatosensory and proprioceptive shortfall are overcome. Nevertheless, people lacking taste usually do not eat and may eventually die if there is no medical intervention. This is often observed in patients who undergo radiotherapy especially head and neck cancer patients [5], they usually encounter a radiation-induced loss of taste that might be whole and often interferes with eating [6]. For their feeding, they usually need the placement of a chronic nasogastric tube

A. D. Cheok and K. Karunanayaka, *Virtual Taste and Smell Technologies for Multisensory Internet and Virtual Reality*, Human-Computer Interaction Series,
https://doi.org/10.1007/978-3-319-73864-2_2

because they are nutritionally compromised already. Notwithstanding, the imperative part of taste in feeding in other species, is that taste plays a part in recognizing and distinguishing hydrocarbons that functions as a pheromonal social communication indication such as in courting and mating. It is still uncertain if taste plays a social communication role in humans similar to olfaction [7, 8].

Taste sensation could be divided into several psychological characteristics: quality, intensity, oral location and timing. These characteristics are, thus, quickly assessed and, inside a given setting, pervaded with some level of positive or negative hedonic value, yummy or yucky [2]. The characteristics of taste are the fundamental subdivisions of the methodology marked sweet, salty, bitter, sour and umami, and these tastes are different from each other. Archetypal models of these tastes are honey, chicken soup, lemon juice, table salt, strong black coffee. The taste intensity is the magnitude of the qualitative sensations, such as weakly salty or strongly bitter. The region in the oral cavity from which a taste sensation emerges is called the location. Unlike smell, taste has a spatial capacity based on its localization that allows humans to localize exact qualities and even to control substances in the mouth [9]. The timing of taste refers to whether taste perception emerge rapidly and if they delayed, i.e. delayed flavor impressions. Timing and spatial prompts are inalienably used to assess stimuli and are to a great extent in charge of why artificial sweeteners seldom taste like sucrose; most artificial sweeteners are localized more to the back of the oral cavity and stay longer than sucrose [1].

However, other factors such as smell [10], texture [11] and temperature also contribute to the perception and flavor of food in the mouth other than primary taste which contribute only partly. We identify food items as familiar or new through flavor which is formed when taste combines with smell. Every basic taste is designated either as aversive or appetitive dependent on the impact the sense have on the bodies, since the sense of taste can perceive both useful and hurtful things. For example, sweet taste aids in recognizing foods rich in energy while bitter taste aids as an indication of poisons [12, 13]. Taste perception among humans starts to decline around 50 years of age on account of loss of tongue papillae and a general reduction in salivation production [14]. Similarly, not all mammals share a similar taste senses: a few rodents could taste starch, cats cannot taste sweet, and numerous other carnivores as well as hyenas, dolphins, and ocean lions, have lost the capacity to detect up to four of their inherited five taste senses [15].

The next Sect. 2.2 will explore the gustatory system, the taste buds and discussion on mechanism of taste in humans. Next, in Sect. 2.3 will be the description of the five basic tastes followed by Sect. 2.4 discussing different types of taste receptors and genes. The importance of taste especially to human will be discussed in Sect. 2.5 and how humans can adapt to taste is discussed in Sect. 2.6, followed finally by taste disorders in Sect. 2.7.

2.2 The Gustatory System

The gustatory system is the sensory system responsible for the perception of taste and flavor. The gustatory system in humans is made up of taste cells in the mouth, several cranial nerves, and the gustatory cortex. The human gustatory system (taste) plays a key role in enhancing ones everyday life experiences through memory and emotions. The memory of taste lasts longer than verbal memory [16]. Taste aids us in decision regarding what to eat and impact how we effectively digest these foods. Flavour from food activates the salivary gland to produce digestive juice which aid in digestion and assimilation of food [2]. Taste sensation is distributed across a large area in the oral cavity and pharynx, even though some specific structures are known to facilitate this sense, but how less specific endings play their part is still not known. Nevertheless, examining individual response of fungiform papillae in man, established the base of the papilla to be the sensitive part [17].

The gustatory system is made up of taste receptor cells which are incorporated in structures called papillae found in various parts of the tongue as shown in Fig. 2.1. Three kinds of papillae are involved in taste, they include fungiform papillae located on the anterior two-thirds of the tongue, the foliate papillae situated on the posterior edge of the tongue, and circumvallate papillae located on the posterior third, and is surrounded by a groove [18]. And a fourth type called filiform papillae, that do not contain taste buds [19]. Each fungiform papilla contains one to five taste buds, while each foliate or circumvallate papilla contains hundreds of taste buds.

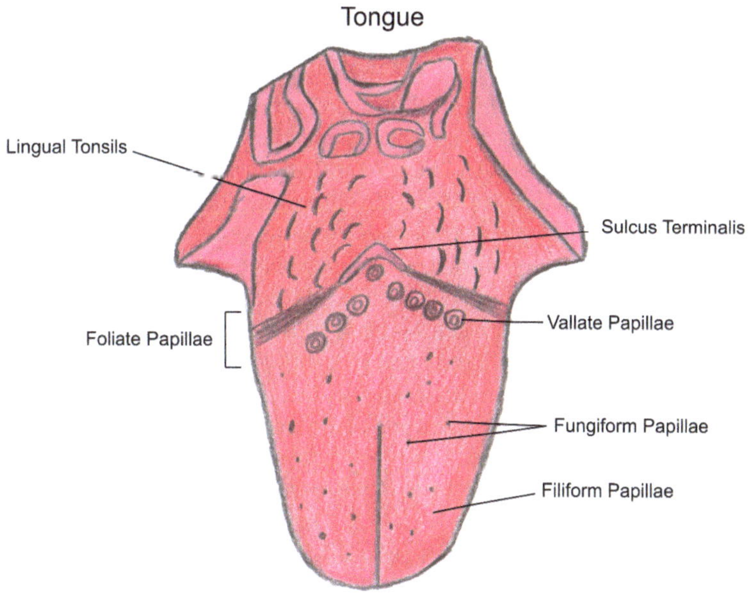

Fig. 2.1 Schematic diagram showing the distribution of papillae along the surface of the human tongue

Besides the papillae, taste receptors are also found in the palate and early parts of the digestive system like the larynx and upper esophagus. There are three cranial nerves that stimulate the tongue namely; the vagus nerve, glossopharyngeal nerve, and the facial nerve. The taste buds of the anterior two-third of the tongue are innervated by the gustatory fibers that travel in a branch of the facial nerve (VII) called the chorda tympani. The taste buds of the posterior third of the tongue are innervated by gustatory fibers that travel in the lingual branch of the glossopharyngeal nerve (IX) [20].

2.2.1 Taste Bud

The taste bud is a microscopic 'rosebud'-shaped structure that comprises between 60–120 cells [21–23]. The receptor cells involved in basic taste signal transduction are in direct contact with the solutions of the oral cavity through microvilli at the apical end of the cells [18]. The microvilli contact the oral solutions through a small (approx. 20 μm) opening in the epithelium called the taste pore. That lies at the tip of each bud. Chemical stimuli are constrained in their flow past taste cells at the taste pore by tight junctions connecting the cells in contact with the pore; generally, only small ions may pass tight junctions [24, 25]. Although, many papillae contain zero buds, adult human fungiform papillae comprise of about four or five taste buds on average [25, 26]. The foliate papillae contain several buds on either side of the epithelial walls that comprise each foliate groove [18]. Likewise, around each vallate papillae, many taste buds line the papillary sidewall of the 'mote' [18]. In contrast, most papillae on the tongue small conical-shaped are filiform papillae, which does not contain taste buds [27].

These appear to serve the purpose of making the lingual surface mechanically rough, which facilitates food and beverage manipulation and may also improve lingual somatosensory function. The larger and less thick fungiform papillae might be seen among the smaller more abundant filiform papillae by simple direct visual examination of the anterior dorsal surface of the tongue [27]. There are four primary kinds of cells inside a taste bud, nonetheless these might be further subdivided histochemically. The cell types were historically labeled dark, light, intermediate and basal cells in view of their electron-dense appearance, shape, and position in an electron microscope image of a taste bud [22, 28]. Located at the base of the taste bud are the small round basal cells. The other three cell types were stretched cells elongating from the basal to the apical end of the taste bud and appearing dark, light, and intermediate [29, 30]. These cell types today are known as type I, II, and III cells, respectively [31]. Both type I and II cells possess microvilli with those of type II cells shorter than those of type I [28, 32]. Only in type II cells, are found most main signal transduction components such as receptors and effector enzymes [33]. Some has presumed that type II cells are the canonical taste receptor cell because of this observation. Though, most synapses with main afferent axons are on type III cells, a considerable number of which are recognized as being serotonergic. However, synapses have likewise been periodically identified in type I and II cells in mice. In taste bud organization, rats, rabbits, and mice seem to have remarkable species differences. The exact configuration of human taste buds remains unresolved [29, 30, 34].

A crucial question is how information flows from type II cells, where the transduction elements is, to type III cells, where most synapses with primary afferents occur [35]. Bud cells communicate with one another neurochemically, though could be electrically coupled through gap junctions. Taste bud cells are known to have serotonin and serotonin receptors, ATP, ADP and their P2X and P2Y receptors, glutamate and its mGluR1, mGluR4, ionotropic receptors, and nitric oxide (NO), this gas (NO) allows cells to interact with each other freely. These are all probable candidates for cell-to-cell chemical interaction within the taste bud, so that main receptor cells lacking neural synapses can interact with taste bud cells with neural synapses [5, 36]. The problem arises as to how to organize the cells inside a bud of a similar chemical sensitivity, since the taste bud contains receptor cells specific for various kinds of chemicals. This could be achieved intragemmally pan-bud by the coordination of the chemical identity of what is released with the appropriate receptors on target cells within the bud. For instance, glutamate may be a bitter compound intercellular signal in the bud, while serotonin may be a sweet compound intercellular signaler [36].

Taste bud cells are different from olfactory receptor cells because they are not neurons, but are nevertheless, similar to olfactory receptor cells in that they have a short life span of approximately 10 days. This is because taste buds contain lived cells that are eradicated same time like the surrounding epithelial cells and are continuously replaced throughout the life of the bud [37]. Within the taste bud, basal cells were thought to give rise to the three elongated cell types and to be the sole progenitors of the elongated cells, however this is not necessarily the case [21, 38, 39]. The stem cells that give rise to taste bud cells reside outside the bud, near its base in the stratum germinativum, and consistently migrate into the bud to produce new cells [38, 39]. Furthermore, the cell types are not all in dissimilar stages along a single cells life cycle. Rather, the stem cells that give rise to various elongate bud cells are of different origins; that is, the elongate cells seem to have different lineages [40]. The role that each bud cell plays remains uncertain. For instance, although type I cells have many long microvilli reaching out into the taste pore and might often have synapses with neurons in some species, it is not clear if they are involved directly in signal transduction. Some have estimated that they play a secretory role for the taste pore and additionally may even serve a glia-like function for the bud [41].

2.2.2 *How Humans Taste*

In humans and mammals, the oral cavity contains receptors for taste, known as taste buds. This microscopic, onion-shaped structures are located in the surface of the tongue and on small protrusions, known as papillae [42]. Taste buds are also found on the roof of the oral cavity, the pharynx and the larynx, but in smaller numbers. All qualities of taste can be elicited from all the regions of the tongue that contain taste buds. Taste researchers have known for many years that the taste map is wrong. The tongue has four types of papillae as discussed earlier including filiform (with no taste buds), circumvallate, foliate and fungiform. The fungiform, foliate, and circumvallate papillae contain many taste buds. Taste buds are stimulated by chemicals in the food we eat and therefore referred to as chemoreceptors. Chemical substances in food and drinks dissolve in the saliva and enter the taste pores, small openings that lead to the interior of the taste bud. Each taste bud contains 30–50 taste receptor cells (TRCs), these TRCs project microvilli known as taste hairs to the surface of the taste bud, where they form the taste pore; the taste pore is in contact with the fluid portion of food within the mouth. Taste molecules from food are believed to bind to hair-like cilia that project from the top of the taste cells. Figure 2.2 shows different type of cells involved in gustatory process. The plasma membranes of taste hairs contain clusters of protein molecules that serve as receptors. These receptors bind to food molecules dissolved in water [43].

After the tastants bind to the cell, taste transduction is slightly different for each of the basic tastes. For salty and sour tastes, the chemicals that produce them act directly via ion channels, however, those producing sweet, bitter and umami tastes bind to

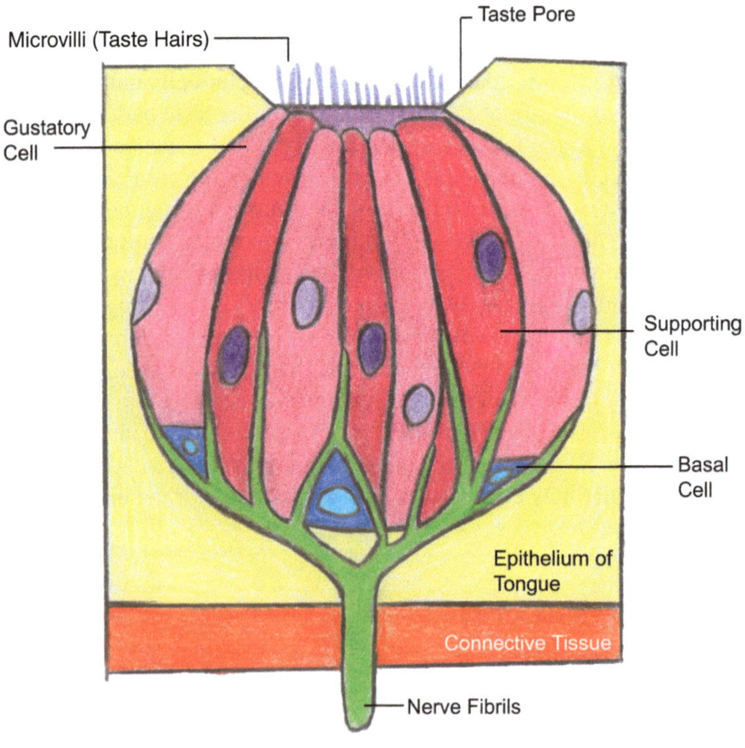

Fig. 2.2 Taste cells involved in the gustatory process

surface receptors that activate a series of signals to cells that eventually results in the opening and closing of ion channels. The concentration of positive ions increases by the opening of the ion channels inside the taste cells. This depolarization then causes the taste cells to release tiny packets of chemical signals called neurotransmitters, which provoke neurons associated to the taste cells to spread electrical messages to

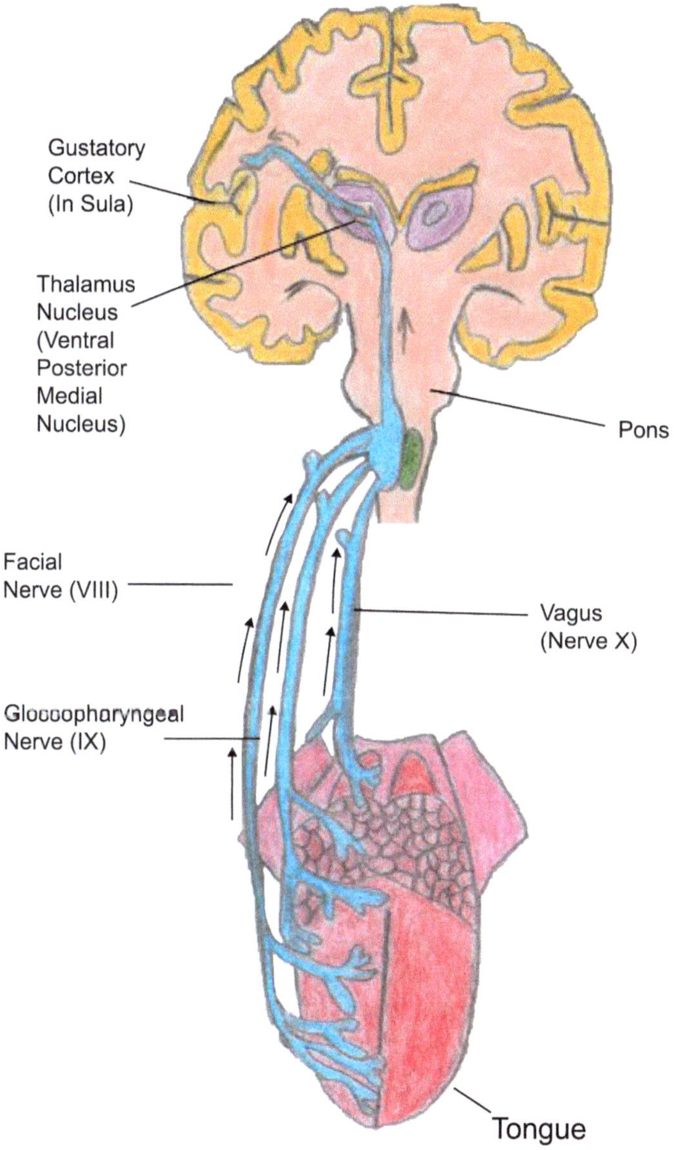

Fig. 2.3 Ascending Gustatory Pathway from tongue to the brain

the brain. This in turn, stimulates the receptor cells, which then stimulate the dendrites of the sensory nerves wrapped around the receptors cells. Impulse from the taste buds are then transmitted to the brain [2]. The process is as shown diagrammatically in Fig. 2.3.

2.3 Basic Tastes

Humans can discriminate among thousands of taste sensations [44]. Sense of taste enables humans to differentiate amongst safe and unsafe food, and to measure the nutritional value of food. The tongue is enclosed with thousands of small bumps called papillae, which can be seen with the naked eye. Inside each papilla are several taste buds, which enables substances to be perceived as taste after been dissolved by digestive enzymes contained in the saliva [45]. Bitter foods are usually thought as unpleasant, just as sweet, sour, salty, and meaty tasting foods commonly produce a pleasant sensation. The five precise tastes received by taste receptors are sweet, sour, salty, bitter, and umami, which means 'delicious' in Japanese and in English is sometimes translated as 'savory'. Western physiologists and psychologists as of the early twentieth century believed there were four basic tastes: sweet, sour, salt, and bitter. Umami was not identified at that time [46] but now many authorities identify it as the fifth taste.

2.3.1 Sweet Taste

Sweet taste is one of the basic tastes usually perceived when eating foods abundant with sugars, regarded most times as a pleasant experience, except possibly in excess. Fructose is sweeter than glucose and sucrose. Because of this, production of sugar syrups with the sweetness and certain other properties of sucrose is made possible starting from starch [47]. Sweet taste is normally viewed as a pleasant sensation, and is delivered mainly by the presence of sugars and some proteins. Sweet taste is frequently associated with ketones and aldehydes, which consists of a carbonyl group. Sweet is identified by a variety of G protein coupled receptors coupled to the G protein gustducin located on the taste buds. For the brain to register sweet taste, no less than two unique variations of the "sweet taste receptors" must be activated [48]. Compounds the brain perceives as sweet are therefore compounds that can bind with varying bond strength to two dissimilar sweet taste receptors. Sweet taste in humans, is stimulated by sugars and an extensive variety of different chemicals which are mostly described as sweeteners. However, in non-human mammals, some of these compounds share similar taste quality and stimulate appetitive consummatory behavior. Sugars are appetizing not exclusively to many vertebrate species, but also to some insects [48].

2.3.2 *Sour Taste*

Sour taste is the taste that is produced by acids and appear attractive to human and animals at low concentration such as in fruits (lemon, grape, orange, tamarind, and sometimes melon) and candy (most contain citric acid). In the UK and US, reports shows that children enjoy sour flavors more compared to the adults [49]. This is believed to be related to children's food habits and preferences. Wine usually has a sour tinge to its flavor, milk also if not properly kept can spoil and has a sour taste. However, it can evoke natural rejection response if it is from food spoiled by acid-producing micro-organisms or unripe fruits that serve as inverse indicator of sugar content. How sour a substance is, is evaluated in respect to dilute hydrochloric acid (HCl), which has a sour taste index of 1. On the other hand, tartaric acid has a sour taste index of 0.7, citric acid has an index of 0.46 and carbonic acid has an index of 0.06 [50, 51]. Sour taste is stimulated by acids such as citric acid or HCl. Protons are the essential stimulus, although carboxylic acids taste sourer than inorganic acids on a constant pH, indicating a role for the anion. Compounds with a high pH (bases) evoke sensations described as irritating, or astringent; this might in part be due to non-gustatory (somatosensory) factors [52].

2.3.3 *Salty Taste*

Salt is an important part to the human diet and improves the flavor of foods. Salty taste is generated mainly by the presence of sodium ions. Similarly, salty taste can be perceived from other ions of the alkali metals groups, however the further from sodium, the less salty the perception is. In humans, a variety of salts, with the most effective stimulus of this sensation being NaCl, stimulates salt taste [50, 51]. The salty quality of sodium salts is influenced by anions and, might add bitterness, sourness, or even sweetness. Similarly, a lot of non-sodium salts, such as KCl, as well have a salty feature beside other qualitative quality. However, in rodents the qualitative taste perception that one could call saltiness comes precisely from the sodium (or lithium) cation. There is a high premium for finding sodium in the environment as a result of the inability to store adequate of this required electrolyte in the body in few omnivores and herbivores [50–52].

2.3.4 *Bitter Taste*

Bitter taste happens to be the most sensitive of the tastes, and many detect it as displeasing or unpleasant but occasionally it is required and its deliberately added through several bittering agents [53]. Mostly, people have strong undesirable responses to bitter substances, this can also signify toxicity, though present in healthful foods like vegetables. Because of this, the capacity to sense bitterness probable played an essential role in human evolution [54]. Bitter taste is assumed by scientists to have evolved as a defence mechanism to identify possible dangerous toxins in plants. Studies shows that in natural foods, TAS2R38, a variant of bitter taste receptor, can perceive glucosinolates compounds with possible dangerous physiological actions [55].

Many years ago, studies show that chimpanzees, like humans, differ in taste sensitivity to the bitter compound phenylthiocarbamide (PTC). Bitter tastes sensitivity offers a significant means for animals to relate with their environment. It allows animals to control their intake of toxins via permitting the recognition of several toxic compounds in food, mainly noxious compounds produced by plants as a means of protection against herbivores. Ingestion of toxic food source or compounds that may lead to sickness or death is also monitored by bitter taste sensation [56].

2.3.5 Umami Taste

Umami is the fifth basic taste quality mainly stimulated by L-glutamate, majorly in the form of monosodium glutamate in the diet [52]. Umami being an appetitive taste [57], it is depicted as meaty or savory. It is found in many fermented and aged food, and can be tasted in soy sauce and cheese. Also, umami is present in foods such as beans, grains and tomatoes. The biological significance of this basic taste, discovered about 100 years ago, is high, comparable possibly to that of sweet taste [58]. Whether amino acids such as glutamate characterize a primary taste stimulus, or whether umami taste may be derivative of the other taste perceptions is still a matter of debate. Umami generates a strong response in humans only in the context of other flavor owing that it is a 'helper' taste. This might be because free amino acids hardly appear single in nature.

However, in proper situations like in savory foods and meats, amino acids like glutamate are greatly required by humans because it indicates the presence of protein in food. Umami-tasting compounds comprised of some L-amino acids (e.g. glutamate and aspartate), tricholomic, purine 5'-ribonucleotides, gallic and succinic acids, theogallin, theanine and several peptides. In addition to umami taste, most of these compounds have other chemosensory components. For instance, the anion (L-glutamate) in monosodium glutamate (MSG), provokes an umami taste, and the cation (Na^+) gives a salty taste. Comparing tastes of equimolar solutions of MSG and NaCl is an effective way to experience umami. Both solutions have saltiness, which is attributed to sodium, nevertheless MSG also has another taste component called umami which is not found in NaCl solutions [52].

2.3.6 Other Taste Qualities

Beside the basic qualities of taste discussed above, there are quite a few other taste qualities that might not be less essential rather are less understood. Rodents, might have a polysaccharide or starch taste that is different from normal sugar taste. The existence of a fat taste is logical, though still hotly debated, given that the taste system perceives other macronutrients. For animals, it is uncertain whether the fats they sense in the month is because of tactile and olfactory signals only or in together with fat taste. A likely fat receptor CD36 present in the taste buds [59] has been recognized [60], and research has shown it binds long chain fatty acids [61]. For centuries, metallic taste has already been described, but this different sensation may be a combination of several taste qualities with somatosensory inputs from ion-induced currents in the tongue.

Another taste quality discovered is carbonation. It is surprising, taste perception of carbonation might be the least understood aspect of CO2 detection. In the psychophysical literature, the gustatory qualities of CO2 are somewhat controversial; while a few studies support a role of taste in sensing carbonation while others describe CO2 as entirely tasteless to humans and entirely a somatosensory stimulus [62]. Some physiological studies support the existence of a taste response to CO2, though none could connect this to perception of a specific type of taste. These data suggest that in addition to somatosensory detection, carbonation in the mouth could be sensed through one or more gustatory labeled lines, with carbonic anhydrase activity playing some role in the transduction mechanism [63].

2.4 Receptors and Genetic Composition of Taste Qualities

In mammals, taste buds usually contains between 50–100 tightly packed taste-receptor cells (TRCs), representing all five basic qualities: sweet, sour, bitter, salty and umami. Notably, mature taste cells have life spans of only 5–20 days and, consequently, are constantly replenished by differentiation of taste stem cells [64]. Based on the taste buds cytoplasmic electron density, morphology and cytohistochemical profiles, several types of taste receptor cell have been recognized. The real receptor proteins are located in the apical membranes of a subset of the taste bud cells. The apical membranes bulge via a break, called the taste pore, in the stratified squamous epithelium lining the oral cavity. This lining and the tight junctions between taste bud cells offers protection from the possibly dangerous chemicals positioned in the mouth. Bitter, and umami tasting stimuli bind to seven-transmembrane spanning G-protein coupled receptors, whereas salty and sour stimuli are said to communicate directly with ion channels [52]. Remarkably, a considerable number of the taste G-coupled receptors and their critical intermediate transduction components are found mainly in Type II cells, which do not have conventional synapses with gustatory neural fibers. The neural synapses occur mainly with Type III cells. Such discoveries lend support to the perspective of the taste bud as a processing unit.

The taste feedback of an individual in genetic terms, represents a phenotype. A constant quantitative scale is used in measuring several taste phenotypes and are thus considered quantitative traits. One of the aim of genetic study is to demonstrate if the phenotypical difference among individuals has a genetic component. Genes with allelic variations that underlie variety of quantitative characters exist in chromosomal areas called quantitative trait loci (QTL) [65]. Quantitative characters that rely upon various genetic as well as environmental factors are observed as complex characters. Most often, taste perception phenotypes are most often complex traits. Usually, the genetic examination involves related individuals, for example twins, or families, or crosses among inbred strains of mice or rats. Linkage study comprises assessment of relationship amongst chromosomal markers, for instance single nucleotide polymorphisms (SNPs) and phenotypes in these populations [65]. Notwithstanding, gene studies helps to clarify variations in taste perception as well as explaining specie differences. Several taste-related genes are orthologs in humans and animals owing to evolutionary relatedness amongst species. This explains the use of mice and rats as laboratory animals, and as model organisms to study genetics of taste sensation. In this section, we will be discussing basically taste receptors and genetics in human and mice [65].

2.4.1 Receptors and Genes Responsible for Sweet Taste

Sugars, artificial sweeteners, Amino acids, and some sweet tasting proteins are known by heteromers of the T1R family of receptor proteins; T1R1, T1R2 and T1R3. L-amino acids binds the T1R1 + T1R3 heteromer, and its stimulation is accelerated by the presence of 5 ribonucleotides such as inosine monophosphate. In humans, the T1R1 + T1R3 receptor appears to identify only L-glutamate and L-aspartate. A few other amino acid receptors have similarly been proposed to be receptors of umami ligands, for example, the splice variant of the metabotropic glutamate receptor subtype 4 (mGluR4) Fig. 2.4. The perception of sweet taste seems to arise in huge part from stimulation of the T1R2 + T1R3 heteromer. For both L-amino acid and sweetener receptors, gene knockout data from mice have persuasively shown the need of the T1R heteromeric receptors for typical taste sensation, though each protein might can form low affinity homomeric receptors. Whether the T1R receptors only are adequate for amino acid and sugar taste sensation has been more difficult to decide [52, 65, 66].

In several animals, appetitive responses to sweet taste stimuli are inherent, however they are likewise frequently modulated by environment and rely upon genetic features. This makes sweet taste sensation a complex trait, which most often has polygenic inheritance. Genes contribute to variations (i.e. T1R-dependent and T1R-independent variations) in sweet taste sensation within and between species. Differences in sweet taste is a function of the variations in sequences of the T1R gene, how-

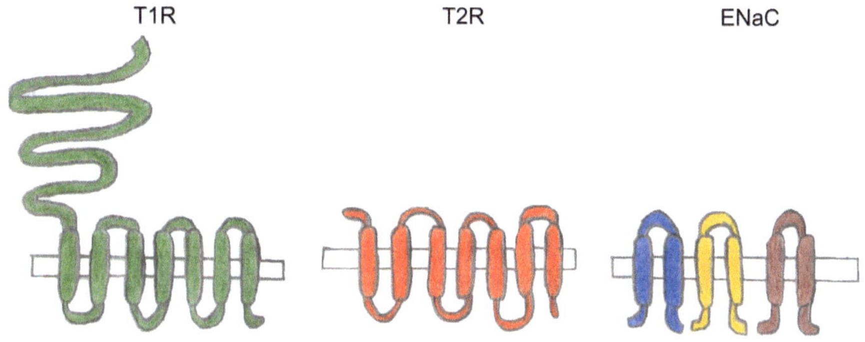

Fig. 2.4 Known taste receptors. The T1R (a; sweet and umami) and T2R (b; bitter) proteins are G protein-coupled receptors, while ENaC (c; salty) is an ion channel

ever, most of this difference is T1R-independent. The T1R-independent difference might be credited to the genes associated with sweet taste transduction, transmission and central processing. Even though individual difference in human sweet taste perception often described as a 'sweet tooth' was known for a long time, nevertheless, understanding of the genetic basis for this difference began recently to emerge [65]. The TAS1R2 and TAS1R3 genes in human have many polymorphisms. Some of these variants are connected with how much sugar is consumed [67] or how sensitive the sucrose taste is [68]. Studies on candidate gene association show that both peripheral and central mechanisms are involved in T1R-independent genetic variation in sweet taste inclination. In the GNAT3 gene are SNPs encoding the taste exact GA protein subunit gustducin showed in taste bud cells in the tongue, are connected with human sucrose perception [68]. Amino acid coding variants in the dopamine D2 receptor (DRD2) and glucose transporter type 2 (GLUT2) genes (both expressed in the brain) are related with constant consumption of sugars by humans [67].

2.4.2 *Receptors and Genes Responsible for Bitter Taste*

In the plant and animal world, bitter stimuli are abundant and thus acts on the largest set of oral taste receptors. Before now, humans have 25 putatively functional bitter receptor genes coding for G-coupled receptors, known collectively as the T2R family. The T2Rs differ essentially in their binding profile some seems precise for only a limited compound, though others are triggered by entire chemical classes. Presently, more than half of these receptors remain orphans with respect to their ligand binding features [66]. A couple of interesting features of T2R expression have implications for taste coding. Some taste receptors cells seem to respond selectively to various bitter tasting ligands, while others seem to respond to both quinine and sucrose. It is also worth noting that T1R and T2Rs have been found to be expressed in the gastrointestinal tract, where they might play a part in enteric nutrient signaling [52, 65, 66].

The genetic origin of bitter taste in both humans and mice has been studied widely [65]. Individuals show what is most often substantial variation in bitter taste sensitivity, both in terms of absolute threshold and suprathreshold estimation of intensity. The difference in sensitivity between tasters and non-tasters to phenylthiocarbamide (PTC) compound created an important early model for what appeared to be a case of single gene inheritance in humans [69]. Sensitivity to other bitter compounds has also been found to be heritable, examples include quinine and sucrose octaacetate (SOA) [70]. Based on dietary importance, it is gradually clear that much of this difference is connected to TAS2R polymorphisms selected for TAS2R38. For example, a polymorphic bitter taste receptor gene is accountable for about 85% of the difference in PTC sensitivity amongst human subjects [69].

2.4.3 Receptors and Genes Responsible for Salty Taste

In the mouth, the simplest taste receptor is sodium chloride (salt) receptor. Epithelial sodium channel (ENaC), an amiloride-sensitive ion channel which facilitates salty taste especially in rodents. This channel is extremely selective for Na+ (and Li+), giving rodents a remarkable capacity to sense and identify sodium salts irrespective of the anion. Another form of the Transient Receptor Potential V1 (TRPV1) channel has been found as a nonselective cation channel implicated with responses to distinct types of salts. Amiloride, in humans does not seem to modify the saltiness of NaCl, but rather reduces its extremely weak sour taste. This proposes species differences in the functions these ion channels play in expressing salt taste [52, 65, 66].

Individuals categorized as PROP 'supertasters' with heightened sensitivity to the bitter compound propylthiouracil tend to report larger saltiness in salt solutions and specific foods [71]. Though, a current heritability research found no evidence of genetic control of saltiness [72]. Inherited strain variations have been accounted for NaCl preference/aversion, NaCl thresholds and amiloride[1] sensitivity of chorda tympani[2] responses to NaCl; the last was connected to allelic[3] difference in the a-ENaC subunit [73]. ENaC channels assume a vital part in salt taste for various vertebrate species, going from rodents to dogs and to primates.

---

[1] Amiloride is a synthetic pyrazine derivative inhibiting Sodium (Na) reabsorption through the Na channel.

[2] Chorda tympani is a branch of the facial nerve that originates from the taste buds in front of the tongue, runs through the middle ear and carries taste information to the brain.

[3] Allelic is any of two or more variants of a gene that have the same relative position on homologous chromosomes and are responsible for alternative characteristics.

2.4.4 *Receptors and Genes Responsible for Sour Taste*

Several candidate sour taste receptors have been proposed, nevertheless none has been absolutely proven [74]. Lately, a member of the PKD1 family of TRP channels, the polycystic kidney disease-like ion channel PKD2L1, which is shown in the apical region of a subset of taste bud cells, has been implicated as a component in taste transduction involving acid stimuli function as sour taste receptors [75]. Though, targeted mutations in these genes have no or only modest outcome on taste responses to acids. Several species of animal senses and typically avoid sour taste. Heritable differences in sour taste were reported in human twins [72] and among inherited mouse strains [76]. These genetic studies might assist with identification of the still elusive sour taste receptor through the positional cloning approach, which enabled identification of T1R and T2R receptors. To link difference in sour taste ability with possible polymorphisms in candidate sour taste receptors, further studies are required [52, 77].

2.4.5 *Receptors and Genes Responsible for Umami Taste*

In humans, a heterodimer[4] of T1R1 and T1R3 proteins functions as an umami taste receptor, but in rodents it is more broadly tuned to respond to L-amino acids [65]. For umami or glutamate taste, quite a few other molecules have been anticipated as candidate mammalian taste receptors. Humans vary in sensation of glutamate taste [78], and current studies show that polymorphisms[5] in the TAS1R1, TAS1R3 and mGluR1 genes partially explain this difference [73, 79, 80]. The TAS1R1 and TAS1R3 heterodimer receptor functions as the umami receptor, responding to L-amino acid binding L-glutamate [81]. The umami taste is most frequently associated with the food additive monosodium glutamate (MSG) and can be improved through the binding of inosine monophosphate (IMP) and guanosine monophosphate (GMP) molecules [82].

2.5 The Importance of Taste

Taste, especially for omnivorous species is a vital sense owing that the potential choice of foods, their difference in nutrient content, and the risks of accidental ingestion of poison increases with the variety and complexity of the feeding plan. Compared to species with highly specialized diets, for example, the leaf-eating koalas and giant pandas. They have less dietary choices to make and are faced with less

[4]Heterodimer is a micromolecular complex formed by two different proteins.

[5]Polymorphism is a discontinuous genetic variation resulting in the occurrence of several different forms of individual among the members of a single species.

risks from poisons than do omnivores. Therefore, their gustatory systems seem to have decreased. Clearly because of the colossal reduction in their selected food types comparative to other bears, giant pandas have lost an amino acid taste receptor gene TAS1R1 [83]. However, carnivorous mammals by contrast have retained the amino acid taste receptor, but lost numerous other taste receptors from their genome. All cats for instance, have lost their recognized sweet taste receptor gene, TAS1R2 [84]. Though we could not certainly know why these genes have been lost in these lineages, possibly the receptors were no longer beneficial or essential for survival.

Aquatic carnivorous mammals, such as sea lions, have much more taste receptor pseudogenes and seem to have lost many taste receptors, possibly because most of their prey are swallowed whole and would not be tasted [85]. Thus, many species seem to have lost a few or all their taste receptors since they do not need the exact nutrient detectors. An implicit question is, what essential functions does taste serve for humans? Have humans retained a variety of functional taste receptors because of the desire to taste? First, taste sensory inputs impact our thinking, deciding, and conduct toward sampled foods, both consciously and unconsciously, to guide ingestion. Secondly, taste inputs impact our physiology and the metabolic processing and signaling of nutrients and poisons as soon as it is ingested. The former is involved with determining what foods enter our body and the latter with how these nutrients are taken care of as soon as they entered. Together these two functions help create our food preferences and feeding habits that maintain and sustain us through life and empower our species to reproduce.

Generally, through taste eating is considered a pleasurable experience. Digestion starts by taste and smelling food because they trigger the salivary glands and digestive fluids. Digestion and assimilation would have been difficult without these sensations. Information about the food we eat are provided by these senses, making them vital to our health. We can detect if food is fresh through its aroma. Spoiled food can also be detected by smelling it or taking a tiny bit of it. Humans depend on these senses of smell and taste to keep away from foods that may be harmful. Humans tend to lose appetite in their food if they are unable to taste or smell their food, and this results to poor nutritional status and weight loss. For health and well-being of humans, the capacity to taste and smell our food and environment is very important [2].

2.6 Taste Sensitivity and Adaptation

2.6.1 Taste Sensitivity

The sensitivity with which an individual perceive different tastes and flavor is referred to as taste sensitivity. A person's taste sensitivity is determined by the number of taste buds, trigeminal nerve endings as well as their sensory capacities and how the brain responds to signals from taste buds and trigeminal nerve endings. The number and appearance of taste papillae affects taste sensitivity. Therefore, people with huge number of papillae together with high sensitivity of the nerve ending experiences taste as being very strong. But people with lesser taste buds and low sensitivity do not experience taste as strong. There are basically three types of tasters; the highly, the moderately and mildly sensitive tasters [86].

2.6.1.1 The Role of Saliva in Taste Sensitivity

The parotid, submandibular, sublingual (major glands) as well as labial, lingual, buccal and palatal (minor gland) produces saliva. The primary excretory channel of the parotid organ opens on the buccal mucosa close to the upper molar teeth. Saliva is the principal fluid constituent of the external environment of the taste receptor cells and, as then, can play a part in taste sensitivity [87]. Its main function includes transport of taste substances to and protection of the taste receptor. Saliva acts as a solvent for taste substances in the early process of taste perception, it dissolves taste substances, and latter diffuse to the taste receptor sites. Some salivary constituents chemically interact with taste substances during this process. Taste sensitivity is influenced in several ways by saliva, for example through diffusion, chemical interaction of taste substances, stimulation and protection of taste receptors [87]. These several impacts are realized by the numerous natural and inorganic constituents of saliva. It is notable that there is significant individual difference in the saliva composition. Under different conditions, the constituents of the salivary concentration of different people and in the same individual varies significantly. Taste sensitivity may like-wise fluctuate extensively, because of these disparities in saliva.

2.6.2 Taste Adaptation

Taste adaptation is a reduction in response to taste sensation and this happens when the sensory receptors are exposed to prolonged stimulation. It is one of the central phenomena of taste perception. As a result, receptors lose their ability to respond and develop a reduced sensitivity to the stimulus. Hence, continued exposure specifically causes the brain cells to pay less attention to the stimulus and decreases the reaction to the particular sensation [88]. During eating, the taste buds in the mouth play a key role. The tongue has approximately 2,000–8,000 taste buds divided into five basic tastes: sweet, salty, sour, bitter and umami. When eating, the initial taste of the food is very different and it's identified by the tongue's sensory neurons. As the eating continues, the taste diminishes and does not have the same impact, which is due to sensory adaptation [88]. Adaptation to taste has been known for quite a while. Suffice it to say that if the taste receptors start signaling to the brain the existence of a constant stimulus, the message becomes reduced; this can be regarded as a shift in the zero for taste. The attenuation to the taster, is perceived as a steady decrease in intensity of taste frequently resulting in the taste really vanishing (complete adaptation). This attenuation or adaptation seems to be because of a particular inhibition of the taste receptor sites rather than the exhaustion of some receptive substance in the cell. It can be thought of as the volume switch being gradually turned down for the taste apparatus. Moreover, different types of receptor sites on the taste membrane seem to undergo the process of adaptation independently. Taste psychophysicists use this to investigate whether adaptation to one stimulus affects the taste intensity of another stimulus and hence whether they share common receptor sites. Such adaptation impacts everyday eating, are typically less vigorous. In the mouth, the food is moved to different locations mixing with saliva and several food stimuli are repeatedly mixed together [89, 90]. There are 2 types of adaptation effects; in one, thresholds are elevated (cross adaptation) and in the other the thresholds are dropped (cross enhancement).

2.6.2.1 Cross Adaptation

Situations in which adaptation to one substance increases the threshold for another substance therefore decreasing its sensitivity is referred to as cross adaptation. For instance, drinking vinegar or lemonade on fries frequently, the drink will not taste sour due to reduced sensitivity. Because cross adaptation is precisely to a specific taste quality; if one is adapted to a sour taste, sensitivity to sour tastes will be reduced, but sensitivity for salty, sweet, or bitter will not be reduced [89, 90].

2.6.2.2 Cross Enhancement

When exposure to one substance reduces the threshold for another substance and leads to an increase in sensitivity is referred to as cross enhancement. For instance, orange juice might taste unpleasantly bitter if you have been eating a candy. If a particular taste is exposed continuously, its absolute threshold might rise until it is higher than the concentration of the adapting solution. Hence adaptation is complete at this point, and no taste experience occurs. But removal of the adapting solution reverses the adaptation process and the threshold adjusts back to its initial value [90].

2.7 Taste Disorders

Taste could be categorized as either quantitative or qualitative, this classification has shown to be valuable in clinical routine. Quantitative gustatory disorders are ageusia, hypogeusia. Ageusia is a complete loss of taste function, which is a rare occurrence, while hypogeusia is a diminished ability to taste sweet, salty, bitter, sour and umami. Isolated losses of any taste quality such as the failure to taste sweet, while bitter, salty and sour could be tasted [91] is very uncommon and has been described [92]. Qualitative gustatory disorders are primarily parageusia, phantogeusia and dysgeusia. Parageusia is an unpleasant taste evoked by the nutritional intake, which is generally missing, while phantogeusia is the most common taste disorder, which is persistent, usually unpleasant taste [93–95]. And dysgeusia describes a condition in which a foul, salty, rancid, or metallic taste sensation will persist in the mouth. While the normal taste function is known as normogesia. Most times, patients trying to find help due to taste impairment end up having an olfactory problem instead of an isolated taste problem [96]. Since taste and flavor are similar in the present language, a reduction in perception of flavor will result to a condition depicted as taste loss. This is predominantly because of the impact of retronasal olfaction on flavor sensation and highlights the significance of psychophysical taste and smell testing. Many studies regrettably, depend only on the patients complaints and for a long time standardized tests have not been available. Meanwhile taste and smell disorders might regularly occur simultaneously [97, 98], each chemosensory modality ought to be assessed independently before any conclusion.

2.7.1 Causes of Taste Disorders

Taste disorder could be as a result of either of the following factors, age, drugs and metal exposure or pesticide exposure, inflammation in the mouth, infection that decreases flow of blood to the tongue and inhibits production of saliva leading to damage of cell receptors. Other factors includes: gastroesophageal reflux (GERD), systemic conditions such as Crohn disease, diabetes, pernicious anemia, head trauma, radiation treatment and surgical procedures [99]. Taste dysfunction could be grouped into two:

- postoperative and posttraumatic gustatory dysfunction
- neurological gustatory dysfunction

Postoperative and traumatic gustatory disorder could be caused by middle ear surgery, tonsillectomy surgery, oncologic surgery and radiation therapy and, dental procedures [100]. From a neurological point of view, gustatory disorders can result from damage at any location of the neural gustatory pathway from the taste buds via the peripheral (facial, glossopharyngeal and vagal nerve) and central nervous system (brainstem, thalamus) to its representation within the cerebral cortex. Gustatory disorder has also been indicated in Parkinsons and Alzheimers disease and leads to reduction in both olfactory and gustatory sensitivity [99]. Ageing can as well be a factor in taste disorder. In the elderly, isolated taste impairment is common but seldom causes serious clinical problems [99].

2.8 Summary

In this chapter, we have discussed the science of taste which is an important aspect of human life. It is the key sensory modality through which we evaluate whether a potential food is good or harmful. The sense of taste is used by humans to record memory as an important part of daily life activities. Currently, in the digital media taste still receives minor attention although being one of the important chemical sense. In this chapter, we have reviewed few aspects of sense of taste including: basic tastes (sweet, bitter, salty, sour, and umami), taste receptors, description of the gustatory system and its make-up, as well as the description of how humans taste. Here, adaptation to a particular type of taste, different types of taste disorders, and how it affects one's general living were discussed.

References

1. Breslin PA, Huang L (2006) Human taste: peripheral anatomy, tastetransduction, and coding. In: Taste and smell, vol 63. Karger Publishers, pp. 152–190
2. Breslin PA (2013) An evolutionary perspective on food and human taste. Current Biol 23(9):R409–R418
3. Mattes RD (2011) Accumulating evidence supports a taste component for free fatty acids in humans. Phys Behav 104(4):624–631
4. Roper SD (2013) Taste buds as peripheral chemosensory processors. In: Seminars in cell & developmental biology, vol 24. Elsevier, pp 71–79
5. Finger TE, Danilova V, Barrows J, Bartel DL, Vigers AJ, Stone L, Hellekant G, Kinnamon SC (2005) ATP signaling is crucial for communication from taste buds to gustatory nerves. Science 310(5753):1495–1499
6. Yee KK, Li Y, Redding KM, Iwatsuki K, Margolskee RF, Jiang P (2013) Lgr5-EGFP marks taste bud stem/progenitor cells in posterior tongue. Stem Cells 31(5):992–1000
7. Fausto N, Campbell JS, Riehle KJ (2012) Liver regeneration. J Hepatol 57(3):692–694
8. Cowart BJ, Yokomukai Y, Beauchamp GK (1994) Bitter taste in aging: compound-specific decline in sensitivity. Phys Behav 56(6):1237–1241
9. Grill HJ, Norgren R (1978) Neurological tests and behavioral deficits in chronic thalamic and chronic decerebrate rats. Brain Res 143(2):299–312
10. Steven Dowshen M (2013) What are taste buds
11. Eric H (1998) Smell-the nose knows
12. Lange R (2017) Why do two great tastes sometimes not taste great together? (12)
13. Miller G (2011) Sweet here, salty there: evidence for a taste map in the mammalian brain. Science 333(6047):1213–1213
14. Seidel HM, Stewart RW, Ball JW, Dains JE, Flynn JA, Solomon BS (2010) Mosby's guide to physical examination-e-book. Elsevier Health Sciences
15. Scully SM (2014) The animals that taste only saltiness
16. Annett JM (1996) Olfactory memory: a case study in cognitive psychology. J Psychol 130(3):309–319
17. Murray RG (1971) Ultrastructure of taste receptors. In: Taste. Springer, pp 31–50
18. Smith DV, Margolskee RF (2001) Making sense of taste. Sci Am 284(3):32–39
19. Norton N (2007) Parotid bed and gland
20. Shepherd R, Farleigh C, Land D (1984) Preference and sensitivity to salt taste as determinants of salt-intake. Appetite 5(3):187–197
21. Delay RJ, Roper SD, Kinnamon JC (1986) Ultrastructure of mouse vallate taste buds: II. Cell types and cell lineage. J Comp Neurol 253(2):242–252
22. Kinnamon JC, Taylor BJ, Delay RJ, Roper SD (1985) Ultrastructure of mouse vallate taste buds. I. Taste cells and their associated synapses. J Comp Neurol 235(1):48–60
23. Kinnamon JC, Henzler DM, Royer SM (1993) Hvem ultrastructural analysis of mouse fungiform taste buds, cell types, and associated synapses. Microsc Res Tech 26(2):142–156

24. Jahnke K, Baur P (1979) Freeze-fracture study of taste bud pores in the foliate papillae of the rabbit. Cell Tissue Res 200(2):245–256
25. DeSimone JA, Ye Q, Heck GL (1993) Ion pathways in the taste bud and their significance for transduction. In: Ciba foundation symposium 179-The molecular basis of smell and taste transduction, Wiley Online Library, pp 218–234
26. Miller IJ (1986) Variation in human fungiform taste bud densities among regions and subjects. Anat Rec 216(4):474–482
27. Kobayashi K, Kumakura M, Yoshimura K, Takahashi M, Zeng J, Kageyama I, Kobayashi K, Hama N (2004) Comparative morphological studies on the stereo structure of the lingual papillae of selected primates using scanning electron microscopy. Ann Anat 186(5–6):525–530 (Anatomischer Anzeiger)
28. Pumplin DW, Yu C, Smith DV (1997) Light and dark cells of rat vallate taste buds are morphologically distinct cell types. J Comp Neurol 378(3):389–410
29. Azzali G (1997) Ultrastructure and immunocytochemistry of gustatory cells in man. Ann Anat 179(1):37–44 (Anatomischer Anzeiger)
30. Azzali G, Gennari P, Maffei G, Ferri T (1996) Vallate, foliate and fungiform human papillae gustatory cells. An immunocytochemical and ultrastructural study. Minerva Stomatol 45(9):363–379
31. Farbman AI, Hellekant G, Nelson A (1985) Structure of taste buds in foliate papillae of the rhesus monkey, macaca mulatta. Dev Dyn 172(1):41–56
32. Yee CL, Yang R, Böttger B, Finger TE, Kinnamon JC (2001) Type III cells of rat taste buds: immunohistochemical and ultrastructural studies of neuron-specific enolase, protein gene product 9.5, and serotonin. J Comp Neurol 440(1):97–108
33. Yang R, Tabata S, Crowley HH, Margolskee RF, Kinnamon JC (2000) Ultrastructural localization of gustducin immunoreactivity in microvilli of type II taste cells in the rat. J Comp Neurol 425(1):139–151
34. Paran N, Mattern CF, Henkin RI (1975) Ultrastructure of the taste bud of the human fungiform papilla. Cell Tissue Res 161(1):1–10
35. Kinnamon JC, Sherman TA, Roper SD (1988) Ultrastructure of mouse vallate taste buds: III. Patterns of synaptic connectivity. J Comp Neurol 270(1):1–10
36. Herness S, Zhao FL, Kaya N, Shen T, Lu SG, Cao Y (2005) Communication routes within the taste bud by neurotransmitters and neuropeptides. Chem Senses 30(suppl_1):i37–i38
37. Farbman A (1980) Renewal of taste bud cells in rat circumvallate papillae. Cell Prolif 13(4):349–357
38. Beidler LM, Smallman RL (1965) Renewal of cells within taste buds. J Cell Biol 27(2):263–272
39. Conger AD, Wells MA (1969) Radiation and aging effect on taste structure and function. Radiat Res 37(1):31–49
40. Stone LM, Tan SS, Tam PP, Finger TE (2002) Analysis of cell lineage relationships in taste buds. J Neurosci 22(11):4522–4529
41. Lindemann B (1996) Taste reception. Physiol Rev 76(3):719–766
42. Docherty BA, Alport LJ, Bhatnagar KP, Burrows AM, Smith TD (2010) Tongue morphology in infant and adult bushbabies (otolemur spp.). In: The evolution of exudativory in primates. Springer, pp 257–271
43. Ranasinghe N, Cheok A, Nakatsu R, Do EYL (2013) Simulating the sensation of taste for immersive experiences. In: Proceedings of the 2013 ACM international workshop on Immersive media experiences, ACM, pp 29–34
44. Chiras DD (2012) Human body systems. Jones & Bartlett Publishers
45. Boron WF, Boulpaep EL (2012) Medical physiology, 2e updated edition e-book: with student consult online access. Elsevier Health Sciences
46. Ikeda K (2002) New seasonings. Chem Senses 27(9):847–849
47. Potter NN, Hotchkiss JH (2012) Food science. Springer Science & Business Media
48. Beauchamp GK, Cowart BJ (1987) Development of sweet taste. Sweetness, pp 127–140

49. Liem DG, Mennella JA (2003) Heightened sour preferences during childhood. Chem Senses 28(2):173–180
50. Guyton AC (1991) Textbook of medical physiology, 8th edn. WB Saunders Company, Philadelphia, p 782
51. McLaughlin S, Margolskee RF (1994) The sense of taste. Am Sci 82(6):538–545
52. Breslin PA, Spector AC (2008) Mammalian taste perception. Curr Biol 18(4):R148–R155
53. Scinska A, Koros E, Habrat B, Kukwa A, Kostowski W, Bienkowski P (2000) Bitter and sweet components of ethanol taste in humans. Drug Alcohol Depend 60(2):199–206
54. Campbell MC, Ranciaro A, Zinshteyn D, Rawlings-Goss R, Hirbo J, Thompson S, Woldemeskel D, Froment A, Rucker JB, Omar SA et al (2013) Origin and differential selection of allelic variation at TAS2R16 associated with salicin bitter taste sensitivity in africa. Mol Biol Evol 31(2):288–302
55. Sandell MA, Breslin PA (2006) Variability in a taste-receptor gene determines whether we taste toxins in food
56. Wooding S, Bufe B, Grassi C, Howard MT, Stone AC, Vazquez M, Dunn DM, Meyerhof W, Weiss RB, Bamshad MJ (2006) Independent evolution of bitter-taste sensitivity in humans and chimpanzees. Nature 440(7086):930–934
57. Jaco PT (2009) Why do two great tastes sometimes not taste great together? (May)
58. Lindemann B (2001) Receptors and transduction in taste. Nature 413(6852):219
59. Simons PJ, Kummer JA, Luiken JJ, Boon L (2011) Apical CD36 immunolocalization in human and porcine taste buds from circumvallate and foliate papillae. Acta Histochem 113(8):839–843
60. Laugerette F, Passilly-Degrace P, Patris B, Niot I, Febbraio M, Montmayeur JP, Besnard P (2005) CD36 involvement in orosensory detection of dietary lipids, spontaneous fat preference, and digestive secretions. J Clin Investig 115(11):3177
61. Baillie A, Coburn C, Abumrad N (1996) Reversible binding of long-chain fatty acids to purified fat, the adipose CD36 homolog. J Memb Biol 153(1):75–81
62. Cowart BJ (1998) The addition of CO2 to traditional taste solutions alters taste quality. Chem Senses 23(4):397–402
63. Yarmolinsky D (2014) Mechanisms for taste sensation of carbonation. Columbia University
64. Lee H, Macpherson LJ, Parada CA, Zuker CS, Ryba NJ (2017) Rewiring the taste system. Nature 548(7667):330–333
65. Bachmanov AA, Boughter JD, Genetics of taste perception. eLS
66. Chandrashekar J, Hoon MA, Ryba NJ, Zuker CS (2006) The receptors and cells for mammalian taste. Nature 444(7117):288
67. Eny KM, Wolever TM, Corey PN, El-Sohemy A (2010) Genetic variation in TAS1R2 (ile191val) is associated with consumption of sugars in overweight and obese individuals in 2 distinct populations. Am J Clin Nutr, 29836 (ajcn)
68. Fushan AA, Simons CT, Slack JP, Manichaikul A, Drayna D (2009) Allelic polymorphism within the TAS1R3 promoter is associated with human taste sensitivity to sucrose. Curr Biol 19(15):1288–1293
69. Kim Uk, Jorgenson E, Coon H, Leppert M, Risch N, Drayna D (2003) Positional cloning of the human quantitative trait locus underlying taste sensitivity to phenylthiocarbamide. Science 299(5610):1221–1225
70. Hansen JL, Reed DR, Wright MJ, Martin NG, Breslin PA (2006) Heritability and genetic covariation of sensitivity to PROP, SOA, quinine HCL, and caffeine. Chem Senses 31(5):403–413
71. Hayes JE, Sullivan BS, Duffy VB (2010) Explaining variability in sodium intake through oral sensory phenotype, salt sensation and liking. Physiol Behav 100(4):369–380
72. Wise PM, Hansen JL, Reed DR, Breslin PA (2007) Twin study of the heritability of recognition thresholds for sour and salty taste. Chem Senses 32(8):749–754
73. Shigemura N, Shirosaki S, Sanematsu K, Yoshida R, Ninomiya Y (2009) Genetic and molecular basis of individual differences in human umami taste perception. PLoS One 4(8):e6717
74. Bachmanov AA, Beauchamp GK (2007) Taste receptor genes. Annu Rev Nutr 27:389–414

75. Ishimaru Y, Inada H, Kubota M, Zhuang H, Tominaga M, Matsunami H (2006) Transient receptor potential family members PKD1L3 and PKD2L1 form a candidate sour taste receptor. Proc Natl Acad Sci 103(33):12569–12574
76. Bachmanov A, Tordoff M, Beauchamp G (2000) Acid acceptance in 28 mouse strains. Chem Senses 25:600
77. Bachmanov A, Bosak N, Lin C, Matsumoto I, Ohmoto M, Reed D, Nelson, T (2014) Genetics of taste receptors. Curr Pharm Des 20(16):2669–2683
78. Lugaz O, Pillias AM, Faurion A (2002) A new specific ageusia: some humans cannot taste l-glutamate. Chem Senses 27(2):105–115
79. Chen QY, Alarcon S, Tharp A, Ahmed OM, Estrella NL, Greene TA, Rucker J, Breslin PA (2009) Perceptual variation in umami taste and polymorphisms in TAS1R taste receptor genes. Am J Clin Nutr 90(3):770S–779S
80. Raliou M, Boucher Y, Wiencis A, Bézirard V, Pernollet JC, Trotier D, Faurion A, Montmayeur JP (2009) TAS1R1-TAS1R3 taste receptor variants in human fungiform papillae. Neurosci Lett 451(3):217–221
81. Zuker CS, Ryba NJ, Nelson GA, Hoon MA, Chandrashekar J, Zhang Y (2008) Mammalian sweet taste receptors. US Patent 7,402,400, 22 July 2008
82. Nelson G, Chandrashekar J, Hoon MA, Feng L, Zhao G, Ryba NJ, Zuker CS (2002) An amino-acid taste receptor. Nature 416(6877):199–202
83. Zhao H, Yang JR, Xu H, Zhang J (2010) Pseudogenization of the umami taste receptor gene TAS1R1 in the giant panda coincided with its dietary switch to bamboo. Mol Biol Evol 27(12):2669–2673
84. Li X, Li W, Wang H, Cao J, Maehashi K, Huang L, Bachmanov AA, Reed DR, Legrand-Defretin V, Beauchamp GK et al (2005) Pseudogenization of a sweet-receptor gene accounts for cats' indifference toward sugar. PLoS Genet 1(1):e3
85. Jiang P, Josue J, Li X, Glaser D, Li W, Brand JG, Margolskee RF, Reed DR, Beauchamp GK (2012) Major taste loss in carnivorous mammals. Proc Natl Acad Sci 109(13):4956–4961
86. Mennella JA, Pepino MY, Duke FF, Reed DR (2010) Age modifies the genotype-phenotype relationship for the bitter receptor TAS2R38. BMC Genet 11(1):60
87. Matsuo R (2000) Role of saliva in the maintenance of taste sensitivity. Crit Rev Oral Biol Med 11(2):216–229
88. Henrichon SJ (2017) Examples of sensory adaptation
89. Breslin P (2001) Human gustation and flavour. Flavour Fragr J 16(6):439–456
90. Wolowich J (2017) Examples of sensory adaptation
91. Henkin R, Shallenberger R (1970) Aglycogeusia: the inability to recognize sweetness and its possible molecular basis. Nature 227(5261):965–966
92. Fox AL (1931) Six in ten tasteblind to bitter chemical. Sci News Lett 9:249
93. Rutherfoord GS, Mathew B (1987) Xanthogranuloma of the choroid plexus of lateral ventricle, presenting with parosmia and parageusia. Br J Neurosurg 1(2):285–288
94. Petzold G, Einhäupl K, Valdueza J (2003) Persistent bitter taste as an initial symptom of amyotrophic lateral sclerosis. J Neurol Neurosurg Psychiatry 74(5):687–688
95. Nocentini U, Giordano A, Castriota-Scanderbeg A, Caltagirone C (2004) Parageusia: an unusual presentation of multiple sclerosis. Eur Neurol 51(2):123–124
96. Deems DA, Doty RL, Settle RG, Moore-Gillon V, Shaman P, Mester AF, Kimmelman CP, Brightman VJ, Snow JB (1991) Smell and taste disorders, a study of 750 patients from the university of pennsylvania smell and taste center. Arch Otolaryngol Head Neck Surg 117(5):519–528
97. Hummel T, Nesztler C, Kallert S, Kobal G, Bende M, Nordin S (2001) Gustatory sensitivity in patients with anosmia. Chem Senses 26:118
98. Probst-Cousin S, Rickert C, Kunde D, Schmid K, Gullotta F (1997) Paraneoplastische limbische enzephalitis. Der Pathol 18(5):406–410
99. Heckmann J, Lang C (2006) Neurological causes of taste disorders. In: Taste and smell, vol 63. Karger Publishers, pp 255–264
100. Landis BN, Lacroix JS (2006) Postoperative/posttraumatic gustatory dysfunction. In: Taste and smell, vol 63. Karger Publishers, pp 242–254

Chapter 3
Science of Olfaction

3.1 Introduction

Olfaction (sense of smell), from evolutionary perspective is one of the most ancient chemical senses [1], however, they remain in many ways the least understood of the sensory modalities. This chapter presents the general review of this very important chemical sense.

The human sense of smell is part of chemosensory system, which helps to discriminate a vast variety of odors and flavors. Olfaction is a fundamental sense for humans, as well as animals, and has played a vital role in the behavior of many animals. Olfaction allows vertebrates and other organisms with olfactory receptors to identify food, mates, predators, and provides both sensual pleasure as well as warnings of danger. For both animals and humans, it is one of the important means by which our environment communicates with us [2, 3]. Perhaps the most significant role of olfaction is to aid animals to distinguish between an extensive variety of alluring and unpleasant objects, and frequently it plays a determining part in specific interaction. Though, it was long considered of secondary importance to senses such as vision or hearing [4, 5].

It is generally believed that human olfaction is limited because of the lesser number of olfaction genes as compared to rodents and dogs. Humans have about 350 functional smell genes while rodents have more than 1100 olfaction genes [6, 7]. On the other hand, [8] in his review shows that humans have complex olfactory bulbs and cortices which provides more sensitive and dynamic abilities for the sense of smell and as a result humans could detect and distinguish at least 1 trillion different smells [9], making humans more sensitive to some odor than rodents and dogs.

Since different individuals respond differently to smells, the physiological and psychological state can be affected by smell through two mechanisms: (a) the intrinsic pharmacological properties of the odor molecule itself and (b) contextual association and memory. Contrary to expectations, research has shown humans adapt to unpleasant smells more quickly than pleasant smells. Awareness of an unpleasant smell is turned off more quickly, this may appear as a bad approach since bad smells

A. D. Cheok and K. Karunanayaka, *Virtual Taste and Smell Technologies for Multisensory Internet and Virtual Reality*, Human-Computer Interaction Series,
https://doi.org/10.1007/978-3-319-73864-2_3

signals to possible danger (bad or spoiled food, poisons, toxins, etc.) [10]. After we have taken in the smell information carried by an unpleasant smell, we either avoid it or, if we think it is harmless for the moment, we adjust to it. And immediately the level of unpleasant smell is altered we became aware of it again [11]. This does not happen with pleasant smells, emphasizing the point that pleasant smells have less biological significance, at least in terms of survival. Yet pleasant smells can create changes in heart rate, respiration rate, blood oxygen, skin resilience and blood pressure [11]. Also, we tend to adapt to smells the longer we get exposed to them. For instance, the longer one uses a particular odorant, at a time the user will not be able to identify the concentration of the smell.

The sense of smell was discovered to be deeply rooted in the limbic system of the brain, an area that governs emotion, behavior and long-term memory. Smells can have a direct and subconscious effect on mood, specific emotions, attitudes, work efficiency, perceived health, emotional memory, and emotionally conditioned behavior [12]. Effects of smell on behavior can occur without conscious awareness [13]. Unlike with other senses, when we smell, our brain reacts before we think about what we are smelling.

For the sake of clarity this chapter has been divided thus: the first section of this chapter provides a general introduction on science of olfaction, Sect. 3.2 will discuss the olfactory system and perception, while the theories of olfaction, chemical interaction as well as process of olfaction will be described in Sect. 3.3. Section 3.4 will highlight different characteristics of smell while importance of olfaction will be discussed in Sects. 3.5 and 3.6 will be a brief discussion of how people adapt to smells, smell dysfunction and effects of olfactory loss.

3.2 Olfactory System Anatomy

The olfactory system (OS) is the part of the sensory system used for smelling (olfaction). Most mammals and reptiles have a key olfactory system and an additional olfactory system called the accessory system. The main olfactory system identifies airborne substances, while the accessory system detects fluid-phase stimuli. The olfactory system as represented in Fig. 3.1 is the most thoroughly studied component of the chemosensory triad and processes information about the identity, concentration, and quality of a wide range of chemical stimuli. These stimuli, called odorants, interact with olfactory receptor neurons in an epithelial sheet, the olfactory epithelium, that lines the interior of the nose [14]. The olfactory system consists of a sensory organ (the olfactory epithelium) and exact olfactory brain areas, the first is the olfactory bulb [14].

The perception of smells presents intriguing and dissimilar problems for the nervous system, which is exclusive to the odorous world. First attribute being that there is no single aspect that connects stimulus to sensation. Vision and hearing are stimulated by foreseeable differences in frequencies of light and sound; touch is stimulated by differences in frequencies of pressure on the skin. Odorant molecules

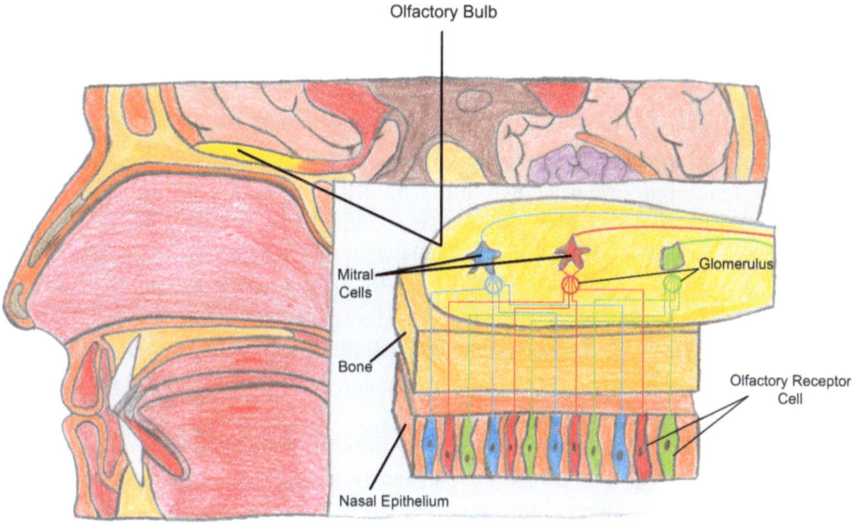

Fig. 3.1 Scheme of the human olfactory system displaying: the olfactory bulb, mitral cells, bone, nasal epithelium, glomerulus, and olfactory receptor cells

have no clear connections with each other apart from that they are odorous, that is, they induce sensations in the olfactory system. The second exceptional characteristics of the olfactory system is that it appears there is no restriction to the quantity of odorous molecules that could be distinguished and described. There are basically a countless quantity of odorous molecules. Touch, vision, and hearing all work inside restricted spectra of light, sound and pressure, unsurprising spectra to which the system have advanced [14].

The question is, in what way, can a system advance to perceive and react to such an open-ended set of stimuli. Nevertheless, the olfactory system has resolved this problem by generating a vast amount of distinct receptor genes. In the mouse genome is about 30,000 genes, olfactory receptors account for more than 1000 [15]. One-thirtieth of the genome is been dedicated to identifying smells. Humans have approximately 900 olfactory receptor gene sequences nevertheless 63% are interrupted with sequences in order to make them non-coding, therefore, they are referred to as 'pseudogenes' [16]. In spite of the fact that people, with their lessen reliance on smell, have approximately 350 dynamic olfactory receptor genes [16]. This implores the question; what is it about the stimulus that necessitates such an enormous investment of genes compared to the visual visual framework which requires just three three genes to identify the color spectrum. Likewise, the auditory system only needs an extra ordinary physical structure called the cochlea, which is designed through genes used for several additional functions in development [14].

3.2.1 The Olfactory Epithelium

The olfactory mucosa is the part of the nasal cavity that its main function is for the detection of odorants. It comprises the olfactory epithelium; a patch of tissue about the size of a postage stamp covered in mucus that lines the nasal cavity and its underlying lamina propria as shown in Fig. 3.2. In humans, the olfactory epithelium measures approximately 1 cm^2 (on each side) and lies on the top of the nasal cavity around 7 cm above and behind the nostrils. It is the part of the olfactory system directly in charge of identifying smell. The olfactory mucosa ostensibly is situated above and back part of the nasal cavity, near the cribriform plate through which the olfactory nerves project to locate the olfactory bulb. The olfactory mucosa in adult is not always adjacent and could be found more anteriorly and inferiorly on the nasal septum and lateral wall [17, 18].

Remarkably, the olfactory epithelium go through a continuous process of neurogenesis in which new neurons are constantly produced through adult life, and this could clarify the discontinuity and spread of the olfactory mucosa [17]. The axons of the sensory neurons exit from the olfactory epithelium and assemble in groups inside the lamina propria to course superiorly and posteriorly. This will cause them to assemble into bigger fascicles that shape the fila olfactoria that navigate the

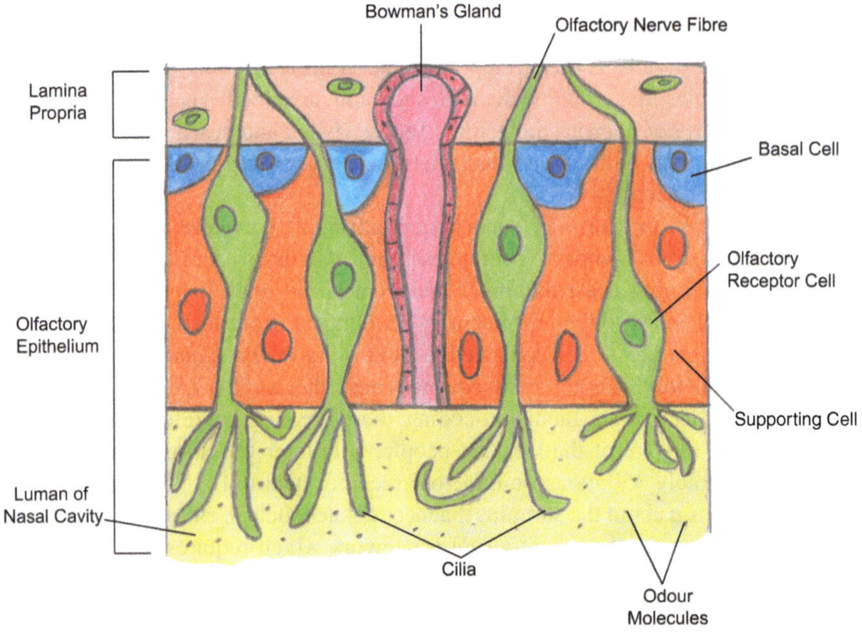

Fig. 3.2 Olfactory epithelium comprises of olfactory receptor cells, which have specialized cilia extensions. Cilia traps odor molecules as they pass across the epithelial surface and information about the molecules are transmitted from the receptors to the olfactory bulb in the brain

skull base via the numerous openings of the cribriform plate to go into the olfactory bulb. In the olfactory bulb, the axons will detach and pass into specialized structures, glomeruli, within which the sensory axons synapse with the mitral cells of the olfactory bulb. Inside the olfactory epithelium, the olfactory sensory neurons reveal only one of the numerous odorant receptor genes [15]. Cells expressing each gene are dispersed apparently 'arbitrarily' inside the olfactory epithelium, subject only to subgroups of these cells being restricted to numerous broad regions inside the nasal cavity [19]. These sensory neurons once they direct their axons to the olfactory bulbs, the glomeruli to which they are constrained are restricted to a single glomerulus on either side of the olfactory bulb [20]. Hence, in mice there are approximately 1800 glomeruli [21] and about 1000 receptor genes [15]. Recent study showed that every glomerulus is innervated by sensory neurons showing just single receptor gene [20]. The receptor gene identity exhibited by each sensory neuron is involved in aiming its axon to an exact glomerulus, even though the receptor gene is essential but not adequate to deliver the axon to a specific glomerulus.

3.2.1.1 Basal Cells

Basal cells are stem cells that are capable of separation and differentiation into either supporting or olfactory cells. They rest near the basal lamina of the olfactory epithelium. Though few of these basal cells divide quickly, a considerable proportion remains relatively dormant and replace olfactory epithelial cells as required. This results to the replacement of the olfactory epithelium every 6–8 weeks [22]. Basal cells can be grouped into two populations based on their cellular and histological characteristics: (i) the horizontal basal cells, and (ii) globose basal cells, which are a heterogeneous population of cells comprising of reserve cells, intensifying progenitor cells, and immediate precursor cells [23].

3.2.1.2 Supporting Cells

Similar to neural glial cells, in the olfactory epithelium the supporting cells are non-neural cells that are situated in the apical layer of the pseudostratified ciliated columnar epithelium. In the olfactory epithelium, the supporting cells are of two types: (i) subternatural cells and (ii) microvillar cells. The function of the sustentacular cells is as metabolic and physical support for the olfactory epithelium. Another class of supporting cells are microvillar cells, they are morphologically and biochemically different from the sustentacular cells, and arise from a basal cell population that expresses c-Kit[1] [24].

[1] C-kit is a type of receptor also known as tyrosine protein kinase.

3.2.1.3 Olfactory Receptor Cells

These are bipolar neurons, each having a thin dendritic rod that consist of specialized cilia which continues from the olfactory vesicle and a long central process that forms the fila olfactoria. The cilia provide the transduction surface for odorous stimuli. The apical poles of these neurons express odorant receptors on non-motile cilia at the ends of the dendritic knob, which extend out into the airspace to interact with odorants. Odorant receptors bind odorants in the airspace, which are made soluble by the serous secretions from olfactory glands located in the lamina propria of the mucosa. The axons of the olfactory sensory neurons assemble to form the olfactory nerve. Once the axons pass through the cribriform plate, they terminate and synapse with the dendrites of mitral cells in the glomeruli of the olfactory bulb.

3.2.2 The Olfactory Bulb

Inferior to the basal frontal lobe lies the olfactory bulb. It is a highly organized structure composed of several distinct layers and synaptic specializations. From outside toward the center of the bulb the layers are differentiated thus: glomerular layer, external plexiform layer, mitral cell layer, internal plexiform layer and granule cell layer. The olfactory bulb is likewise connected with the amygdala, which processes emotions, and furthermore the hippocampus, known for its part in learning. Hence, smell is famous for activating memories and intense feelings.

3.2.2.1 The Anatomy and Physiology of Olfactory Bulb

The cellular structure of the olfactory bulb is said to be well recognized [25]. Sensory information coming in, goes to the mitral and tufted cells that supply the output to the higher olfactory centers. Inside the olfactory bulb, this output is intensely modulated by the interneurons available at numerous anatomical and processing stages. Throughout the glomeruli are numerous types of interneurons, a large number of which are dopaminergic, whose axons and dendrites form part of the complex neuropil inside the glomeruli. Also extending far down the olfactory bulb are the granule cells, which does not contain axons, nevertheless the dendrites connect with mitral cell in more superficial layers, modulating their activity through complementary sets of synapses from granule cell dendrites to mitral and tufted cells, and in the reverse direction, the supposed 'dendrodendritic synapses'.

The olfactory bulb physiology is subjugated by the spatial nature of the input. This is precisely because of the spatial patterning of the sensory axons revealing diverse odorant receptor genes, which when presented in the nose activates diverse patches of glomeruli on the surface of the olfactory bulbs. The activity of the mitral cells is spatially dispersed in a way that odorants are represented in the olfactory bulb by a distributed form of mitral cell activity, since each mitral cell contains a dendrite in one single glomerulus [26–28]. The nature of interneuron connectivity then results to improving of the reaction of the mitral cell both in time and space, with the result of narrowing the reactions of the mitral cells to a lesser amount of odorant molecules in relation to the sensory neurons [27, 28]. Likewise, the mitral cell reaction is narrow in time too [27].

A distinguished characteristics of olfactory bulb anatomy is the convergence of feedback from higher centers whose axons extend onto the interneurons at the granule and periglomerular levels. Together with the impression that cellular models of memory are apparent in the activity of olfactory bulb neurons [29], these features propose that the olfactory bulb is not simply a passive conduit for sensory information, modulated and sharpened by interneuron activity. There is abundant indication that the olfactory bulb physiology is controlled by previous exposure and experience with odorants [30] and prove for feedback superimposed on the olfactory bulb [31]. It is thus possible that the olfactory bulb acts as a filter of some sort that matches expected patterns of activity associated with food, mates or predators, for example, with the pattern of sensory nerve activity.

3.2.2.2 Human Olfactory Bulb Size

The size of the olfactory bulb is comparative small in primates like humans compared to the rest of the brain, constituting approximately 0.01% of the human brain by volume in comparison to 2% of the mouse brain. Nevertheless, the absolute olfactory bulb size is moderately large, quite bigger compared to that of mouse and rat. Therefore, it is a natural question whether the olfactory bulb ought to be seen in relative or absolute terms [32].

Brain structure studies across species showed that the size of any given region in the brain is proportional to the overall size of the brain. Total size of the brain could explain over 96% of the difference in the size of individual brain regions across mammals [33]. Nevertheless, there is one glaring exemption to this rule; the olfactory bulb size. Size of the bulb is not dependent on the size of most other regions of the brain and constitute for nearly all the remaining difference. This exception is considered by modern evolution scholars to be one of the three principles of brain measuring: different allometric scaling for each structure, high intercorrelation of structure volumes and comparative independence of the olfactory-limbic system from the remaining parts of the brain. Therefore, the near prevalent thought of the olfactory bulb in proportion to the rest of the brain is probably inappropriate [34].

A justifiable reason to evaluate the bulb in proportion to other structures is missing, thus it appears better to evaluate its absolute volume. Age and experience can be a variable factor in the volume of the olfactory bulb. The volume of the olfactory bulbs is about 60 mm^3 in adult humans. In hyposmic patients, the bulbs have been noticed to decrease by about 25% over time, and, in subjects who experienced childhood maltreatment to be 20% smaller [35]. However, in the rat, between 3 and 18 months of age the olfactory bulb doubles in volume (climaxing at around 27 mm^3) as the animal itself becomes physically bigger throughout adulthood, but this is unlikely to be accompanied by a corresponding increment in olfactory capacities. In adult mouse, the olfactory bulb volume varies from 3 to 10 mm^3 across strain and study. Across mammalian species, the relative volume of the olfactory bulb is negatively correspondent with general brain size. Notwithstanding these noticeable variances in volume, there is little support for the view that physically bigger olfactory bulbs predict better olfactory function, irrespective of whether bulb size is measured in absolute or relative terms [8]. Why does the olfactory bulb have approximately constant number of neurons across species? Historically, the relationship between brain size and organism size has been interpreted to reflect the inherently larger information processing needs of larger animals, more muscle fibers to organize, more somatosensory input to interpret, and so forth. But, since the size of the organism does not regulate the smells in its environment or its necessity to perceive olfactory stimuli, this logic appears not to apply to the sense of smell [8].

3.2.3 *How Human Olfactory Structures Differ from Those of Other Mammals*

The human olfactory structure especially the olfactory bulb which is responsible for processing sensory input from the nose is different from other mammals. This maybe because after birth this area ceases from developing new neurons [36]. Regardless of the exceptionally comparable amount of neurons found in the olfactory bulb, the olfactory system in human still has distinguished variation compared to other mammals. Human olfactory bulb is arranged into an average of 5600 glomeruli, compared to the mouse or rat with approximately 1800 and 2400 glomeruli respectively. And each glomerulus obtains information from a subpopulation of sensory neurons that all express a similar smell receptor, thereby create a glomerular map that characterizes smell identity [8].

Therefore, in humans, combining higher number of glomeruli and a lesser amount of functional smell receptor genes, implies that humans might have around 16 olfactory bulb glomeruli that processes information from single smell receptor type compared to approximately 2 in rodents. Absence of the "accessory" olfactory system (AOS) a set of parallel structures in human as well as the vomeronasal organ and accessory olfactory bulb which are present in several other animals may be a distinguishing factor between humans and animals. Initially, the AOS was accepted to be particularly for pheromone recognition, but now it is understood to be a general-purpose system for sensing low-volatility scents in liquid phase. Odor-based interaction amongst organisms from same species can work together with the key and AOSs and appears in species with and without an AOS, this also applies to humans [37].

Examination of carbon-14 in neuronal DNA, according to early study showed that neurogenesis does not take place in the adult human olfactory bulb regardless of being conspicuous in hippocampus and striatum [38]. In comparison with rodents, through the animal's life, adult neurons play an ongoing role in olfactory bulb function. Nevertheless, regardless of the absence of adult neurogenesis, the human olfactory system seems capable of much functional plasticity supported by neurogenesis in rodents. The possibly most significant distinction between human and that of other animals olfactory processing is that humans have much more intricate cortical regions for construing olfactory information. Particularly for the orbitofrontal cortex, that is probably bigger and more complicated in humans than in rodents, and which makes wide contacts to other neocortical regions. These differences could permit the system to incorporate smells into background or semantic networks, or to experience plasticity to maintain function after peripheral damage, or to integrate learned information [39].

3.3 Olfactory Perception

Odor perception is influenced by many factors unique to each individual as well as external environmental factors. The basis of odor perception is the contact between chemical molecules, mainly in the gaseous state and the nose, which can be detected by the olfactory epithelium and stimulation of the olfactory sensory neurons and terminates in higher cerebral centers which, when activated, make us consciously aware of an odor.

3.3.1 Olfactory Receptors

Olfactory receptors also known as odorant receptors. Odor recognition and perception occur as a result of stimulation of olfactory receptor neurons (ORNs), which are situated in the primary sensory organs, i.e. the nasal epithelium in vertebrates. The response profile of each ORN is determined by one functional type of olfactory receptor [40]. The numbers of functional OR types range from a few dozen in insects to approximately 600–1400 in vertebrates [40]. Such a high number of independent information channels give olfaction possibly a much higher dimensionality than other modalities. The olfactory receptor in vertebrate are situated in both cilia and synapses of the olfactory sensory neurons and in the epithelium of the human airway. Olfactory receptors show affinity for a variety of odor molecules, rather than binding to specific ligands and equally a single odorant molecule might bind to a number of olfactory receptors with varying affinities which is subject to physio-chemical properties of molecules like their molecular volumes.

In vertebrates, the olfactory receptors are located in both the cilia and synapses of the olfactory sensory neurons [41] and in the epithelium of the human airway [42]. Near the upper conchae are the olfactory receptors. Humans have approximately 400 functional genes coding for olfactory receptors, and the remaining 600 candidates are pseudogenes [43]. The reason for the large number of different odor receptors is to provide a system for discriminating between as many different odors as possible. Even so, each odor receptor does not detect a single odor. Rather each individual odor receptor is broadly tuned to be activated by a number of similar odorant structures.

3.3.2 *Theories of Olfaction*

In the past years, several theories relating odorant quality to molecular structure have been proposed. Here are different theories surrounding olfactory perception. One of them is the shape theory of smell which proposes that a molecule's smell character is because of its molecular shape, molecular size and functional groups [44]. But the complex connection between the shape of an odorous molecule and its perceived smell character is still unclear. The challenge remains regardless of several studies yet the shape-odor relationship is not being discovered [45]. Differently shaped molecules with similar molecular vibrations have similar smells. On the other hand, the Vibration theory of smell as first proposed by Malcolm Dyson in 1928 suggests that a molecule's odor character is because of its vibrational recurrence in the infrared range.

Another theory is called odotope theory, also known as weak shape theory. This theory explains how olfactory receptors bind to odor molecules. The theory suggests that a mixture of shape elements determines the joining. Another theory of odor originally proposed by [46] and extended by [47] is called the Steric Theory of Odor. He proposed that air borne chemical molecules are smelled when they fit into certain complimentary receptor sites on the olfactory nervous system. Primary odors such as ethereal, camphoraceous, musky, floral, minty, pungent and putrid were suggested by Amoore. The steric theory is appropriate to the possibility that the odorant receptor proteins receive just certain odorants at a particular receptor site [48].

Auffarth [1] in his review, started that odorants are detected and thus encoded by distinct sets of olfactory receptors, and resulting in spike generation by olfactory receptor neurons. His study involves the review of several principles of olfactory receptors which according to him were incompletely understood. They include; the transduction principle of olfactory receptors [49], the odotope or weak shape theory [45], the vibration theory [50]. Another work, molecular vibration sensing by [51] which claim that human olfactory receptors can be distinguished using vibrational theory energy level sensing was challenged by another report by [52]. It was concluded that the proposed vibration theory does not apply to neither of the human olfactory receptors examined. Hence, [44] in his report lay to rest the argument against the vibration theory of smell because of lack of evidence.

3.3.3 *Chemical Interaction*

In the sequence of event of olfaction, the odor molecules attaches to the receptor cells and action potentials are produced in the receptor neurons. When the odorant

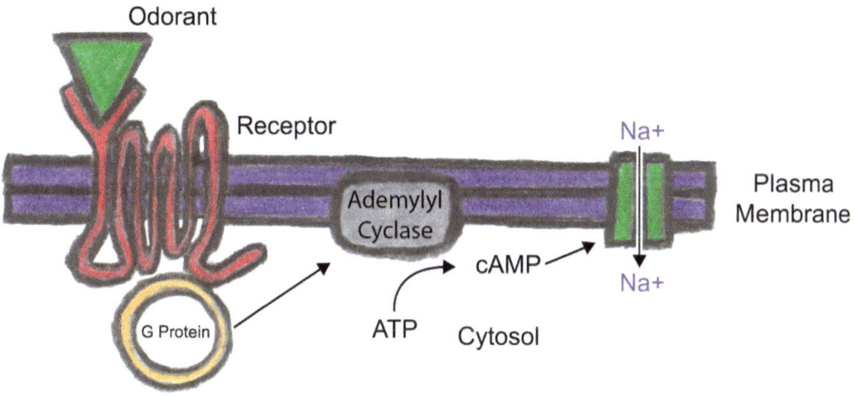

Fig. 3.3 Sequence of event in chemical process of olfaction

interacts with the receptor, numerous changes ensue in the interior of the receptor cell especially in mammals. The chemical events are summarized in Fig. 3.3. Here odorant molecules dissolve in the mucus and then bind to the receptors on the cilia.

When that happens, the receptor protein activates a G-protein which in turn activates the enzyme adenyl cyclase, an enzyme embedded in the plasma membrane of the cilia. Adenyl cyclase then converts ATP to cyclic AMP. The cyclic AMP (cAMP) increased and opens a sodium ion (Na^+) channel and the Na^+ flows into the cell this decreases the potential across the plasma membrane, and trigger Na^+ efflux depolarizing the receptor neurons and then produce an action potential [53]. The action potential is conducted back along the olfactory nerve to the brain and the brain evaluates this and other olfactory signals reaching it and interprets it as a particular odor.

3.3.4 *The Process of Olfaction*

Volatile chemicals can take about two routes to reach the olfactory sensory cells. The routes include orthonasal and retronasal routes. In rethronasal route, the odorant enters the mouth and then diffuses through the nasopharynx to the nasal receptors and then out of the nostrils while in orthonasal route the chemical is carried in the air to the external nostrils and then to the nasal receptors. The air molecules, once inside the nostrils gets to the olfactory epithelium, which is made up of three kinds of cells and contains millions of olfactory receptors that are capable of binding with specific odor molecules [53]. In the olfactory epithelium, odor molecules bind to olfactory receptors which are expressed in olfactory sensory neurons in the nose. Once an odorant binds to the receptor, the olfactory receptors trigger a series of signals to the cells interiors that ultimately results in the opening and closing of ion channels. This increases the concentration of positive ions inside olfactory cells. This

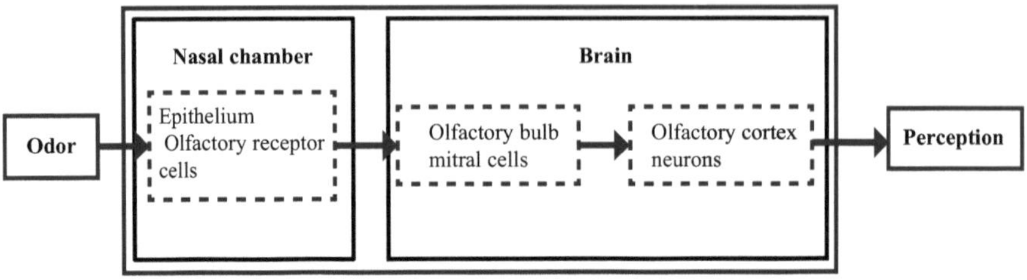

Fig. 3.4 Scheme of the human olfactory process, modified from [54]

depolarization causes the olfactory cells to release tiny packets of chemical signals called neurotransmitters, which initiate a nerve impulse. Odor information is then relayed to many regions throughout the brain [15] which is perceived as smell as illustrated in Fig. 3.4.

3.4 The Characteristics of Sense of Smell

Vision and audition are considered the "higher" senses while touch, taste and smell are the "lower" senses. In the nineteenth century, a distinction was made between the two senses. But are the two types of senses certainly different these days and can grouping them the way they were still be possible? Vision and hearing are involved in such essential human activities such as spatial orientation and communication [55]. Vision and hearing are what drives the great arts e.g. painting, architecture, music, film and photography etc. while touch, taste and olfaction can only demonstration smaller glories e.g. perfumery and cooking and, to some extent, and only in combination with vision. These senses appear very subjective and not as much of universal. However, they are more related to feelings and emotions than to thoughts and decisions. What then are the benefits of having these senses? And what are their characteristics? The senses help us feel at home in our world. Notwithstanding, sense of smell has some distinguishing characteristics different from vision and hearing. The main characteristics of smell will be discussed in the following sections.

3.4.1 The Sense of Smell is a "Hidden" Sense

Olfaction is rarely in the focus of attention unlike vision. We could process much information without being conscious of it in vision. We can be walking on the road and be thinking about other things and not be conscious of vehicles or other road users. But in smell, awareness of odors is exception rather than rule. We perceive smell all the time, it is capable of influencing our moods and the quantity of time we spend in certain locations and perception of other people. A study suggests that odors may have a stronger effect when it is unidentified or the presence unnoticed than when it can be clearly identified [56]. Odor intensity is often suppressed by mixing odors in uneven ratios. Odors do not stop exerting influence on behavior and mood, even when stimuli could no longer be deliberately perceived or are no longer present.

One of the reasons it is difficulty to talk about or describe odors is the fact that humans are rarely conscious of the odor around them. The few abstract words so far for odors like fresh or musty, do not mean the same for everybody. Mostly, odors are designated by the name of their source e.g. strawberry, coffee etc. Based on sensory analysis and from perfumery, it is well known it takes a long training before people can consistently describe olfactory experiences in detail. Furthermore, study has established [57] that odors might be indirectly remembered without awareness of the learning event, and that such memories could be disturbed by clear knowledge of the odor. If sense of smell is certainly a "hidden" sense, therefore in normal everyday life odors may not be intended to be talked about, but just to be experienced with all the amazingly emotional memories they involve.

3.4.2 The Sense of Smell is a "Nominative" Sense

The sense of smell mainly provides us with simple "nominal" data concerning the existence of qualitatively different odors in the surroundings different from vision and audition in which the unconscious measurements of comparative intensity, size and tone play a significant role in informing us about the structure of the world around us. To deliver such nominal information, sense of smell combines good absolute sensitivity for many odors. The nose is still the most sensitive part of the body, having exceptional quality discrimination enabling us to distinguish practically any two odorous compounds offered to us. Identifying odor by name is hard, even though distinguishing a lot of different odors is easy [58]. When two types of odors are presented successively, it is easier to detect whether they are same or different [55]. This shows that in sense of smell, quality discrimination is considered more vital than identification. Contrary in vision, identification is more important than quality discrimination matches are at all times faster than non-matches. At the same time, olfaction has rather poor intensity discrimination. In order to be perceived as different from a given odor concentration, the concentration of that odor has to be raised by about 20% [55]. In vision, an increase of only 2% in brightness suffices to see a difference. Obviously, detection, discrimination and recognition of odors as familiar or unfamiliar are more important than the measurement of intensity gradients or verbal identification of the odor. All these are in accordance with the function of olfaction as a warning system against dangers in the environment and the surrounding air or in food we take into the mouth. Immediate detection of changes in quality is very important, since we cannot afford to stop inhalation and we cannot swallow poisonous food.

3.4.3 The Sense of Smell is a "Near" Sense

The sense of smell has the characteristic feature of being "near" because all the information relating to quality of an object is found in the molecules that are directly in contact with the receptors [55]. On the hand, vision and hearing are termed "far" senses because information about objects are transmitted to the eyes as well as the ear by the intervention of light and sound waves. That the widespread differences in specific olfactory sensitivities also influence the perception of the very complex odor mixtures to which people are exposed in everyday life, will be clear. Thus, it can be concluded that people differ much more in the way in which they perceive odors than in the way they perceive visual objects. These large inter individual differences in olfaction find their origin not only in differences in interpretation as in vision, but also in the sensory bases of the perception itself. That such differences do not harm us, is related to the fact that we do not use odor information in orientation and movement and that we learn to attach (emotional) meaning to odors [55]. Even if a rose, with its mixture of flowery and faecal odors, may smell different to each of us, we all have learned to like the smell and to connect it with love and tenderness.

3.4.4 The Sense of Smell is an "Emotional" Sense

Olfaction plays a vital role in how humans learn to distinguish between foods early in life. Usually, the first food perception were the ones from our mothers and these remain with us for a significant part of our lives. By association we can link odors to emotions and according to studies odor memories have been found to be more emotional than visual memories [59]. Hence, different individual may perceive the same odor differently, depending on their different emotional encounters with the odor in each person. Another study demonstrated that the perceptual responses to smells could easily be influenced by emotional suggestions. In this study, subjects were exposed to known odors and were asked to rate the odor intensities for about 20 min.

3.4.5 The Sense of Smell has a "Special" Memory

In olfaction learning and memory plays a vital part, [60] has given in their review, several number of arguments to back the view that olfactory memory differs from verbal memory. Because olfactory memory is episodic and not semantic in nature, humans tend to remember where and when they have come across an odor before, but in most situations the name of the odor may not be remembered, but most times through guessing we can deduce the name. "When I was small, I smelled this odor in the attic of my grandparent's house. There they kept the apples during winter. It is dried apple smell" [55]. That forms a special memory because it is never forgotten.

Most times when odor and words are presented in a learning situation, odors may be forgotten rapidly compared to words. But when odor is remembered, it stays longer and often forever while words are slowly forgotten entirely. Therefore, inability to remember seems to follow a different time course in olfaction and with verbal stimuli. Furthermore, odors play an important role in making us feel at home in the world. Without any clear awareness, we link emotionally meaningful situations to odor and carry them with us in an implicit memory that makes them typically involuntarily when the odor recurs.

The best means to learn about the actual meaning of the sense of smell and olfactory memory in human life is to listen to people who lost their olfactory sensitivity. These people loss both pleasure of food and drinks because they can not perceive the flavor in either food or drink. Loss of olfaction will affect their emotions and this may eventually lead to depression [61]. Therefore, it is evident that vision and olfaction are different on many important points.

3.5 Human Sense of Smell is Exceptional

It is generally believed that humans have an inferior olfaction compared to other animals such as dogs and rodents. Are humans really inferior to other mammalian species regarding their ability to perceive smells [8]. The answer is "no", because the human olfactory system is exceptional depending on standard of measurement used. When comparing with dogs, dogs might be better in some forms such as discrimination urine smells on fire hydrant, but in discriminating smells of fine wine, humans might be better [8]. However, the primate olfactory system is extremely sensitive to numerous odors when tested correctly, and could employ strong impacts on behavior, physiology, and emotions. Humans whose olfactory systems are intact could perceive almost all volatile chemicals bigger than an atom, in that scientists are now interested in documenting the few odorants that some people cannot perceive [62]. A current study calculated and claim that human could also tell virtually all odors apart, with a projected capability to discriminate more than 1 trillion possible smells [9]. Nevertheless, another study by [63], claim this particular number is extremely sensitive and wrong. Suggesting that the failure is as a result of the mathematical method used to deduce the size of odor space from insufficient experimental sample. However, it is no doubt that the human olfactory system is exceptional at odor discrimination, even better than the presumed 10,000 odors claimed by some studies [8].

Sense of smell strongly influences human behavior. As earlier stated, environmental smell can bring out specific memories and emotions, influence autonomic nervous system activation, shape perceptions of stress and prompt approach and avoidance behavior. The human olfactory system sometimes also plays a major and unconscious role in communication between individuals. Sometimes, after we shake hands with a stranger, we tend to unconsciously smell our hands, signifying an unanticipated olfactory component to this shared social communication [8].

Olfactory communication and perception may be altered by factors such as the developmental stage, sex and age which leads to difference in smelling competence. Sense of smell can also be altered by individual experiences, such as changed odor perception after odor-cued aversive conditioning. Furthermore, the signals from the human olfactory system are always interpreted by the brain in terms of context, anticipation, and previous learning. The human olfaction is however much more essential than we think.

3.6 Olfactory Adaptation and Olfaction Dysfunction

Sensory adaptation, the decrease in response to a continuing stimulus, also happens in olfaction as in other senses. Olfactory adaptation is quick and complete because of a special chemical system that neutralizes the response of OR to smell. In olfaction, smells appear to be 80% less powerful following a couple of minutes of exposure [64]. Because of olfactory adaptation, persons who are wearing excessive fragrance or cologne rarely acknowledge it. The smell is just one-fifth as effective to them as it is to others [64]. A person who spends lots of time working close to an unpleasant smell gets to the point where olfactory adaptation is 100% and he or she cannot smell it at all [64].

Example of olfactory adaptation as experienced by co-author (Sharon Kalu Ufere), while living in an Industrial area of Port Harcourt Nigeria. Where several industries, factories and oil mills were situated close to the residential areas. According to her, initially, she complained of the unpleasant odor that constantly pollutes the air. But as time went by, she got used to the smell that she did not recognize it anymore. Unfortunately, any time a visitor comes around from another town, the visitor will constantly perceive the "terrible smell" in the air. It got to the extent that the family could travel, come back and not realize the difference in smell.

This is different from olfaction dysfunction, whereby sense of smell is either partially or totally lost due to one medical condition or another. About 2 million people in the United States experience some type of olfactory dysfunction. Olfactory dysfunction may be as a result of head trauma, tumors of the anterior cranial fossa, upper respiratory infections and exposure to toxic chemicals or infections [65].

The degree of smell abnormalities is described using the following terms [66]:

- Anosmia: absence of smell sensation
- Dysomia: distortion of smell sensation
- Cacosmia: sensation of an unpleasant or foul smell
- Parosmia: sensation of smell in the absence of appropriate stimulus

3.6.1 *Effect of Olfactory Loss*

Olfactory loss impacts both the health and safety of patient, the quality of life and the vocational life. Patients with olfactory loss are at significant health and safety risk owing to their inability to perceive gas leaks, smoke, spoiled foods and other olfactory-related warning signs that may be existing in the workplace or at home [67]. Among these cited activities, recognition of decayed food happens to be the commonest, followed by detection of leaking gas [68]. Study by [67], found that patients with decreased olfactory function are more likely to experience an olfactory-related dangerous event compared with those with normal olfactory function. Olfactory impairment also affects individual's quality of life [69]. The use of perfumes and cologne decreased and deodorant usage increased. Also, it may affect their hobbies [70]. Younger people are more concerned with their quality of life than older individuals, and women appeared to be more affected powerfully than men. Generally, people with continuous olfactory loss report they were more dissatisfied with their life, than those with improved olfactory function. Persons who suffered olfactory impairment may find it difficult going back to his/her formal vocation. Study showed that patients with anosmia resulting from closed head injury encountered problems when they wanted to go back to their formal jobs [71]. Most demonstrated psychosocial disorders associated with orbital frontal cortex damage, an important brain region for the processing of olfactory information. Olfactory impairment mainly impacts perfumists, florists and cooks. Others such as firefighters and chemists with olfactory impairment may be at risk to themselves and others.

3.7 Summary

The sense of smell help us feel at home in our world. Furthermore, they are more related to feelings and emotions than to thoughts and decisions. Humans smelling ability is said to be weaker compared to other mammalians due to lesser number of olfactory genes. However, humans have complex olfactory bulbs and orbitofrontal cortices which provides them with more sensitive and dynamic abilities and as a result can detect at least 1 trillion different odors. In this chapter, we have answered a few questions regarding the sense of smell, such as the constituents of the olfactory system, how humans perceive smell sensations as well as highlighting the characteristics of smell. The difference between olfactory adaptation and olfactory dysfunction and some negative effects of olfactory loss was discussed as well.

References

1. Auffarth B (2013) Understanding smellthe olfactory stimulus problem. Neurosci Biobehav Rev 37(8):1667–1679
2. Asahina K, Pavlenkovich V, Vosshall LB (2008) The survival advantage of olfaction in a competitive environment. Curr Biol 18(15):1153–1155
3. Croy I, Bojanowski V, Hummel T (2013) Men without a sense of smell exhibit a strongly reduced number of sexual relationships, women exhibit reduced partnership security-a reanalysis of previously published data. Biol Psychol 92(2):292–294
4. Hansson BS, Stensmyr MC (2011) Evolution of insect olfaction. Neuron 72(5):698–711
5. McBride CS (2007) Rapid evolution of smell and taste receptor genes during host specialization in drosophila sechellia. Proc Natl Acad Sci 104(12):4996–5001
6. Shepherd GM (2004) The human sense of smell: are we better than we think? PLoS Biol 2(5):e146
7. Young JM, Friedman C, Williams EM, Ross JA, Tonnes-Priddy L, Trask BJ (2002) Different evolutionary processes shaped the mouse and human olfactory receptor gene families. Hum Mol Genet 11(5):535–546
8. McGann JP (2017) Poor human olfaction is a 19th-century myth. Science 356(6338):eaam7263
9. Bushdid C, Magnasco MO, Vosshall LB, Keller A (2014) Humans can discriminate more than 1 trillion olfactory stimuli. Science 343(6177):1370–1372
10. Moncrieff R (1956) Olfactory adaptation and odour likeness. J Physiol 133(2):301–316
11. Olson CA (2016) Our sense of smell has an effect on pain!
12. Herz RS (2002) Influences of odors on mood and affective cognition. Olfaction Taste Cognit 160:177
13. Selker T, Arroyo E (2002) Interruptions as multimodal outputs: which are the less disruptive. In: IEEE international conference on multimodal interface
14. Mackay-Sim A, Royet JP (2006) Structure and function of the olfactory system. Olfaction Brain:3–27
15. Buck L, Axel R (1991) A novel multigene family may encode odorant receptors: a molecular basis for odor recognition. Cell 65(1):175–187
16. Glusman G, Yanai I, Rubin I, Lancet D (2001) The complete human olfactory subgenome. Genome Res 11(5):685–702
17. Féron F, Perry C, McGrath JJ, Mackay-Sim A (1998) New techniques for biopsy and culture of human olfactory epithelial neurons. Arch Otolaryngol-Head Neck Surg 124(8):861–866
18. Leopold DA, Hummel T, Schwob JE, Hong SC, Knecht M, Kobal G (2000) Anterior distribution of human olfactory epithelium. Laryngoscope 110(3):417–421
19. Ressler KJ, Sullivan SL, Buck LB (1993) A zonal organization of odorant receptor gene expression in the olfactory epithelium. Cell 73(3):597–609
20. Mombaerts P, Wang F, Dulac C, Chao SK, Nemes A, Mendelsohn M, Edmondson J, Axel R (1996) Visualizing an olfactory sensory map. Cell 87(4):675–686
21. Royet J, Souchier C, Jourdan F, Ploye H (1988) Morphometric study of the glomerular population in the mouse olfactory bulb: numerical density and size distribution along the rostrocaudal axis. J Comp Neurol 270(4):559–568
22. Purves D (2001) The olfactory epithelium and olfactory receptor neurons. In: The olfactory epithelium and olfactory receptor neurons
23. Schwob JE, Jang W, Holbrook EH, Lin B, Herrick DB, Peterson JN, Coleman JH (2016) The stem and progenitor cells of the mammalian olfactory epithelium: taking poietic license. J Comp Neurol
24. Goss GM, Chaudhari N, Hare JM, Nwojo R, Seidler B, Saur D, Goldstein BJ (2016) Differentiation potential of individual olfactory c-kit+ progenitors determined via multicolor lineage tracing. Dev Neurobiol 76(3):241–251
25. Mori K, Imamura K, Fujita S, Obata K (1987) Projections of two subclasses of vomeronasal nerve fibers to the accessory olfactory bulb in the rabbit. Neuroscience 20(1):259–278

26. Leon M, Johnson BA (2003) Olfactory coding in the mammalian olfactory bulb. Brain Res Rev 42(1):23–32
27. Lowe G (2003) Electrical signaling in the olfactory bulb. Curr Opin Neurobiol 13(4):476–481
28. Mori K, Nagao H, Yoshihara Y (1999) The olfactory bulb: coding and processing of odor molecule information. Science 286(5440):711–715
29. Wilson RI, Turner GC, Laurent G (2004) Transformation of olfactory representations in the drosophila antennal lobe. Science 303(5656):366–370
30. McLean JH, Harley CW, Darby-King A, Yuan Q (1999) pcreb in the neonate rat olfactory bulb is selectively and transiently increased by odor preference-conditioned training. Learn Mem 6(6):608–618
31. McLean JH, Harley CW (2004) Olfactory learning in the rat pup: a model that may permit visualization of a mammalian memory trace. Neuroreport 15(11):1691–1697
32. Laska M, Genzel D, Wieser A (2005) The number of functional olfactory receptor genes and the relative size of olfactory brain structures are poor predictors of olfactory discrimination performance with enantiomers. Chem Senses 30(2):171–175
33. Hofman MA (1989) On the evolution and geometry of the brain in mammals. Prog Neurobiol 32(2):137–158
34. Nummela S, Pihlström H, Puolamäki K, Fortelius M, Hemilä S, Reuter T (2013) Exploring the mammalian sensory space: co-operations and trade-offs among senses. J Comp Physiol A 199(12):1077–1092
35. Croy I, Negoias S, Symmank A, Schellong J, Joraschky P, Hummel T (2013) Reduced olfactory bulb volume in adults with a history of childhood maltreatment. Chem Senses 38(8):679–684
36. Institutet K (2012) Why humans don't smell as well as other mammals: no new neurons in the human olfactory bulb
37. Spehr M, Spehr J, Ukhanov K, Kelliher K, Leinders-Zufall T, Zufall F (2006) Parallel processing of social signals by the mammalian main and accessory olfactory systems. Cell Mol Life Sci 63(13):1476–1484
38. Bergmann O, Spalding KL, Frisén J (2015) Adult neurogenesis in humans. Cold Spring Harbo Perspect Biol 7(7):a018994
39. Li W, Luxenberg E, Parrish T, Gottfried JA (2006) Learning to smell the roses: experience-dependent neural plasticity in human piriform and orbitofrontal cortices. Neuron 52(6):1097–1108
40. Kaupp UB (2010) Olfactory signalling in vertebrates and insects: differences and commonalities. Nat Rev Neurosci 11(3):188–200
41. Rinaldi A (2007) The scent of life. EMBO Rep 8(7):629–633
42. Gu X, Karp PH, Brody SL, Pierce RA, Welsh MJ, Holtzman MJ, Ben-Shahar Y (2014) Chemosensory functions for pulmonary neuroendocrine cells. Am J Respir Cell Mol Biol 50(3):637–646
43. Gilad Y, Man O, Pääbo S, Lancet D (2003) Human specific loss of olfactory receptor genes. Proc Natl Acad Sci 100(6):3324–3327
44. Vosshall LB (2015) Laying a controversial smell theory to rest. Proc Natl Acad Sci 112(21):6525–6526
45. Sell C (2006) On the unpredictability of odor. Angew Chem Int Ed 45(38):6254–6261
46. Moncrieff RW (1949) What is odor? A new theory. Am Perfumer 54:453
47. Amoore JE (1964) Current status of the steric theory of odor. Ann N Y Acad Sci 116(1):457–476
48. Amoore JE, Johnston JW, Rubin M (1964) The stereochemical theory of odor. Sci Am 210(2):42–49
49. Gelis L, Wolf S, Hatt H, Neuhaus EM, Gerwert K (2012) Prediction of a ligand-binding niche within a human olfactory receptor by combining site-directed mutagenesis with dynamic homology modeling. Angew Chem Int Ed 51(5):1274–1278
50. Malcolm DG (1928) Odour and conetitution among the Pt II. Perfumery Essent Oil Rec 19:88–91
51. Gane S, Georganakis D, Maniati K, Vamvakias M, Ragoussis N, Skoulakis EM, Turin L (2013) Molecular vibration-sensing component in human olfaction. PloS One 8(1):e55780

52. Block E, Jang S, Matsunami H, Sekharan S, Dethier B, Ertem MZ, Gundala S, Pan Y, Li S, Li Z et al (2015) Implausibility of the vibrational theory of olfaction. Proc Natl Acad Sci 112(21):E2766–E2774
53. Leffingwell JC et al (2002) Olfaction–update no. 5. Leffingwell Rep 2(1):1–34
54. Davide F, Holmberg M, Lundström I (2001) 12 virtual olfactory interfaces: electronic noses and olfactory displays. In: Riva G, Davide F (eds) Communications through virtual technology: idnetity community and technology in the internet age. IOS Press, Amsterdam
55. Köster EP (2002) The specific characteristics of the sense of smell. Olfaction Taste Cogn:27–43
56. Degel J, Köster EP (1998) Implicit memory for odors: a possible method for observation. Percept Mot Skills 86(3):943–952
57. Degel J, Köster EP (1999) Odors: implicit memory and performance effects. Chem Senses 24(3):317–325
58. Cain WS (1979) To know with the nose: keys to odor identification. Science 203(4379):467–470
59. Alexander M (2001) How theories of motivation apply to olfactory aromatherapy. Int J Aromather 10(3–4):135–151
60. Herz RS, Eich E (1995) Commentary and envoi. Mem Odors:159–175
61. Castle P, Van Toller S, Milligan G (2000) The effect of odour priming on cortical EEG and visual ERP responses. Int J Psychophysiol 36(2):123–131
62. Schaal B, Porter RH (1991) microsmatic humans revisited: the generation and perception of chemical signals. Adv Study Behav 20:135–199
63. Meister M (2015) On the dimensionality of odor space. Elife 4:e07865
64. Russell A, Dewey P (2007) Olfactory adaptation
65. Costanzo RM, Zasler ND (1992) Epidemiology and pathophysiology of olfactory and gustatory dysfunction in head trauma. J Head Trauma Rehabil 7(1):15–24
66. Levy LM, Degnan AJ, Sethi I, Henkin RI (2013) Anatomic olfactory structural abnormalities in congenital smell loss: magnetic resonance imaging evaluation of olfactory bulb, groove, sulcal, and hippocampal morphology. J Comput Assist Tomogr 37(5):650–657
67. Santos DV, Reiter ER, DiNardo LJ, Costanzo RM (2004) Hazardous events associated with impaired olfactory function. Arch Otolaryngol-Head Neck Surg 130(3):317–319
68. Miwa T, Furukawa M, Tsukatani T, Costanzo RM, DiNardo LJ, Reiter ER (2001) Impact of olfactory impairment on quality of life and disability. Arch Otolaryngol-Head Neck Surg 127(5):497–503
69. Reiter E, Costanzo R (2003) The overlooked impact of olfactory loss: safety, quality of life and disability issues. Chem Sense 6:2–4
70. Temmel AF, Quint C, Schickinger-Fischer B, Klimek L, Stoller E, Hummel T (2002) Characteristics of olfactory disorders in relation to major causes of olfactory loss. Arch Otolaryngol-Head Neck Surg 128(6):635–641
71. Varney NR (1988) Prognostic significance of anosmia in patients with closed-head trauma. J Clin Exp Neuropsychol 10(2):250–254

Chapter 4
Electric Taste

Abstract To pursue the next stage of the internet, humans should not only communicate emotions with visual, audio, and tactile stimuli but also with taste (gustation). Humans will want to share these stimuli collectively as an experience digitally, like they currently do with the visual and audio media on the internet. We want to propose the idea of gustation creating and experiencing digital representation of taste sensation. However, a new methodology is first needed to digitally stimulate the sense of taste to enable internet communication of the sense. We propose an electrical tongue stimulation device, which the user places in their mouth to produce taste sensations. This technique operates by inducing weak electric signals by changing frequency and Pulse Width Modulation (PWM) produced by the circuitry. Several experiments were carried out to evaluate this approach of digitizing taste sensations. Results from experiments show a new user experience through digital taste stimuli.

4.1 Introduction

When Alexander Graham Bell invented the telephone, people could finally listen to music without being physically present at a concert. Later in that century, Paul Nipkow made a device that transmitted visual images which led to the development of the television [1]. Virtual haptic sensation has recently been becoming more accurate in research, and more common in commercial applications such as game controllers. However, a controllable and non-chemical actuation of taste has yet to be achieved. Aside from the enjoyment of food, taste has an advantage over all the other senses in that the neurological interaction between them and our emotions and memories are direct, as both are located in the limbic system of the brain [2].

When we taste foods, our brain reacts to chemical based tastes due to the excitation initiated by a receptor protein in the cell membrane. By changing its conformation when it forms a complex with a stimulant. Mathematical models of the excitation for chemical stimulation of the taste receptors have been proposed [3]. To our knowledge no such analytical models have been proposed for electrical taste. Currently, there are two hypotheses known for the mechanism of electrical taste. The electrolyte-chemical hypothesis proposes that the anode creates an acid solution by electrolysis and the

A. D. Cheok and K. Karunanayaka, *Virtual Taste and Smell Technologies for Multisensory Internet and Virtual Reality*, Human-Computer Interaction Series,
https://doi.org/10.1007/978-3-319-73864-2_4

surplus of positive hydrogen ions then stimulates the taste cells specific for the sour quantity [4]. Previously strong evidence is presented for the idea that hydrogen ions, resulting from the dissociation of food acids, enter sour-specific taste cells that inform the brain about the offensive material. The brain can then decide what to do, often to pull a “sour face [5]”. The direct effect hypothesis, which claims that electrical taste is the result of an action of the current on either the nerve fibers of taste receptor cells [6]. Only the sour-sensitive taste receptor cells are thought to have low apical resistance, which permits the current to enter the cells [7].

There is conflicting research results as previous scientific research has so far been unable to determine precisely which taste qualities are elicited by electrical stimulation. The prevailing theory of electrical stimulated taste is that weak electrical current delivered to the surface of the tongue elicits a sour taste experience [8]. However, recent results have shown that electrogustronomy relates to an ability to appreciate all the five tastes: sour, sweet, salt,bitter and umami. The recent research by [4, 9] showed that electrogustrometry correlated with all five taste qualities. The researchers found that all tastes could be perceived by electrical stimulation. They proposed that there is a direct depolarizing mode of action of the sensoryneural tongue aspects. Previous works have shown correspondence in the taste qualities elicited by electrical and chemical stimulation. Von Békésy [10] used electrical as well as chemical stimulation of single papillae. It was found that using electrical stimulation, a sensation corresponding to one and only one of the primary tastes could be stimulated.

Preliminary experiments have shown that correlations exist between the amounts of current applied and the taste sensation generated [11]. Also, other experiment showed that certain tastes could be artificially stimulated using electrical current stimulation on the human tongue. These sensations were stimulated using an electrical taste device with different experimentally determined PWM. As of now we can use our digital taste interface device to produce some basic taste sensations. We use a combination of electrical signals with electrodes to produce taste sensation on the tip of the tongue. Figure 4.1 illutrates the idea how to develop this electrical taste device. Various trials of this parameters produces different taste sensations. The electrical stimulation module provides square wave pulses to the silver electrode with diverse PWM where it controls the percentage of the duty cycle. In this study, the duty cycle was varied from 0 to 100%. The initial experimental results suggested that sour and salty are the main sensations that could be evoked.

4.2 The Taste Technology Devices

In humans, sense of taste is one of the chemical senses used every day in relating to one another. In the mouth (especially on the tongue) are chemically stimulated receptors responsible for detecting diverse taste sensations. Hence, one or more chemical substances in the mouth are involved in stimulating the sensation of taste. In Human-Computer Interaction (HCI), chemical stimulation of taste has been used to develop interactive systems. Hence, this section discusses few studies with chemical based and non-chemical stimulation methodologies. There are possibility of using non-chemical stimulation methods to stimulate taste sensations digitally. However, the reported studies are conducted mainly in the medical domain (with controlled environments), are invasive, or only in the experimental stage. Therefore, to achieve electrical and magnetic stimulation methods as a means of actuating the sensations of taste, this research will achieve research breakthroughs in controllability, accuracy, and robustness.

4.2.1 *Chemical-Based Approach for Taste Actuation*

There are quite a few studies conducted based on the chemical stimulation of taste. In Human Computer Interaction (HCI), chemical generation of taste sensation has been utilized to develop new systems, though using chemicals in an interactive system to stimulate taste perception is complicated. The following examples of taste stimulation using chemical approach are discussed.

The Food Simulator uses chemical and mechanical linkages to simulate food-chewing sensations by providing flavoring chemicals, biting force, chewing sound, and vibration to the user [12]. The mechanical part of the device comprises primarily of a vibration sensor, vibration motor, and the linkages. While the part inside the

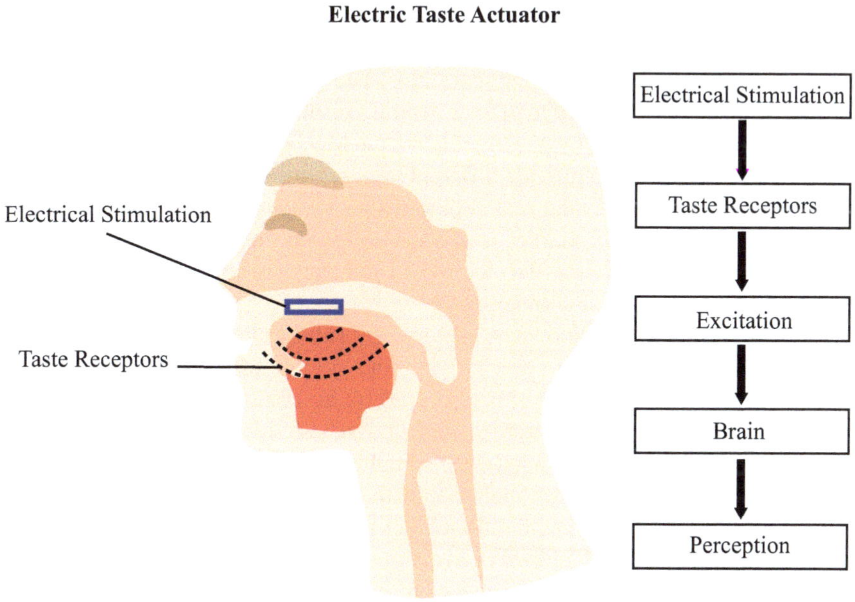

Fig. 4.1 Approach to digital taste stimulation

mouth consist of a rubber cover, meant to prevent a user's bite and the motor provides suitable resistance to the mouth along with chewing sound and chemicals. In this study, the researchers studied the cross sensory interactions of taste with sound, texture, and force [13].

Another taste device using chemicals to stimulate the sense of taste is called TasteScreen [14]. This device is attached to the top of the user's computer screen which holds 20 different chemical flavoring cartridges to mix and spray toward the screen. By licking the computer screen, the user can then taste the released taste.

Another example is Virtual Cocoon which sprays chemicals into the mouth of the wearer to generate a different type of taste sensation [15]. The system stimulates other senses (vision, touch, smell, and audition) as well and not only sense of taste. To produce different flavor, a container of chemicals is connected to a tube which sprays into the users mouth and nose. It uses numerous ranges of chemical to stimulate smell and taste senses, therefore the system is unmovable. As stated by the researchers, the Virtual Cocoon headset may perhaps mimic all five senses and make a virtual world as convincing as real life. However, scientists are still working to make this possible, as to date, Virtual Reality devices have not been capable of stimulating concurrently all five senses with a high degree of realism [16].

Furthermore, as of late there are a few investigations that have shed some light into virtual taste system. For instance, [17] illustrates a pseudo gustatory show in view of the virtual shade of a genuine drink. They utilized a wireless LED (Light Emitting Diode) module connected to the base of a straightforward plastic cup, along these lines to super impose the virtual color of the drink. Outcomes of their experiment demonstrate that distinct colors provoke users to decipher different flavor of a similar drink.

Also, the Tag Candy and Meta cookie systems [18] utilize augmented reality based methods to produce different sensations. The Tag Candy utilizes vibration and hearing via bone conductivity to deliver several sensations as a user enjoys a regular lollipop tied to the system. On the other hand, the Meta Cookie system utilizes smell and visuals information to provide different taste sensations to the user while eating a normal cookie. To cover the real cookie with a virtual cookie in the system, the printed augmented reality marker is used. Also, based on the user's choice, smell information is sent to the user, thus to produce different sensations although the user consumes the same regular cookie in real.

And finally, the device known as gustometer [19], a battery-powered instrument used to deliver a determined concentration and volume of a taste stimulus to the tongue over a specified period. It is also used for assessing the sense of taste. It is made up of two electrodes which are to be placed on both sides of the tongue at various taste centers and that provides a galvanic stimulus resulting in taste sensation. These gustometers could be used on animals to train them to respond to various fluid stimuli differently and to measure unconditioned licking behavior to stimuli presented for short durations to receive a reward or avoid punishment [20].

Audio, visual, and haptic fields are well researched. We believe the next challenge is virtual taste using techniques that are currently unattainable using chemical based methods. Digital taste is one such technique of actuating taste with electric stimulation. This has advantages over chemical stimulation because it is more scalable and taste can be transmitted directly over the internet.

We believe through this research we can eventually digitally actuate all main tastes (bitterness, saltiness, sourness, sweetness, umami) using digital means to actuate and transmit the taste information. It could have long term implications in the future of many disciplines. By using digital taste, we will be able to create new ways to create content and new knowledge of food that can be shared, learned, and created in new interesting ways previously unobtainable with current methods.

4.2.2 Non Chemical-Based Approach for Taste

The scientific knowledge for stimulating the human sense of taste with non-chemical approaches is still in its early stage [13]. One of the early scientists, Alessandro Volta, known for the invention of electric cell and discovery of Voltage, studied the sensory effects of electrical stimulation on human senses particularly for sight, taste and touch. He positioned two coins, made from different metals on the two sides of his tongue (up and down) and linked them via a wire. He said during the experiment, he felt a salty sensation [21] (Fig. 4.2).

Basically, there are numerous indications of stimulating taste sensations via electrical and thermal stimulation in electrophysiology. In a study by [23], electrical stimulation of a single human tongue papillae using a silver wire, providing responses for salty, bitter and sour tastes. Authors used both negative and positive electrical pulses with a frequency range of 50 to 800 Hz. The results provided effective responses for the sour taste (22.2%) and some small responses for the bitter (3.8%) and salty (1.8%) sensations. Lawless presented metallic taste generation from electrical and chemical stimulation [11]. They observed the similarities and differences of stimulation with metals, electrical stimulation, and solutions of divalent salts and ferrous sulphate and investigated sensations that occurred across oral locations using electrical stimulation and different metal anodes and cathodes.

Nakamura shows research using electricity for augmented gustation [24]. They apply electric current through isotonic drinks and juicy foods to change the taste perception of them by pulsing voltage and amplitude input. In his later research, Nakamura propose a saltiness enhancer by applying weak cathodal current for a short time when the user eats or drinks [25]. However other perceptions such as sourness and bitterness was also perceived. Hence he conduct another experiment to compare the intensity of the fundamental tastes and a metallic taste over three phases: before application of current, during application of current, and after the release of current [26]. The user feels that after release of the cathodal current, are more salty than natural foods.

In another research [27] alter the taste of soup by stimulating tounge electrically. They experiment on anodal and cathodal stimulation, using electrode pairs attached to a silver spoon. Result shows that anodal stimulation using this system amplifies acidity, saltiness and taste strength. Sakurai in his work demonstrate cathodal direct

Fig. 4.2 Alessandro Volta [22]

current stimulation to produce saltiness and umami perception [9]. Besides, their experiment shows that the magnitude of the taste suppression is linearly affected by the amplitude of stimulation current. Furthermore, to evaluate the taste recognition thresholds of patients with taste dysfunction, a clinical tool electrogustatory is utilized to electrically stimulate the human tongue. The electrogustometer makes use of direct current with stainless steel electrodes to estimate the threshold of pleasure on patient's tongue [28].

Philips Electronics [29] on the other hand, has a patent on a system to generate taste sensations utilizing electrical stimulation. To determine user's taste preferences, they built a tongue apparatus, which can measure the saliva flow in relation to the stimuli. There are other studies on tongue based interactive systems majorly for physically disable people which generally uses the movements of the user's tongue as an alternative input methodology for computers [30, 31].

Finally, digital lollipop [32], a work of the authors, an electronic device that produces virtual tastes by stimulating the human tongue with electric currents. The device can produce four of the main tastes: bitter, salty, sour and sweet. The system can manipulate the properties of electric currents to formulate different stimuli. Through a sliver electrode, the devices produce alternating current signals stimulating the tongue's taste receptors to mimic the main taste components. To simulate food, it generates small varying amounts of heat as well. The digital lollipop can ultimately be of great importance to Alzheimer's patients by helping them to "either enhance or suppress certain senses" and people with diabetes might as well be allowed to experience sweet taste without increasing their blood sugar levels.

4.3 Methodology

We proposed a new device to artificially induce taste sensations, and technology can be applied to various domains. Our method focused on the accuracy and repeatability of the taste actuation systems, which will enable user to operate this device in an independent manner.

To generate taste sensations digitally, we constructed our own device that can deliver electric current in a controllable and safe manner to the users tongue. This device is shown in Fig. 4.3, where user can place the apparatus on top of the surface of his/her tongue. Controller circuit can generate electric pulses based on the user's input and outputted to the tongue membrane where the tongue makes physical contact with the device. The current travels across the tongue which excites the taste cells that signal the brain that a taste is being sensed. We were interested in the robustness of the system, specifically the sensory adoption, taste controllability through a population, and the comfort while using this system.

We created the PCB using Eagle Software and Fig. 4.4 shows the circuit diagram of the device. It consists of an Arduino Pro mini which receive input from a PC terminal via USB connection. The Arduino can output frequency and PWM as programmed. We use this feature to produce square waves of varying frequencies and PWM to control the taste interface. The user feels this current simulation with their tongue and the result produces several scale of taste sensations.We use a magnitude range of 20–200 uF and a frequency range of 50 Hz–1 KHz.

Since the measured impedance is different in everyones tongues, we created an additional part of the circuit that provides a constant current source using an operational amplifier and an NPN transistor. The current output is delivered to the load (silver tongue electrodes), which are placed on the top and bottom of the tongue.

We mounted the electrodes to the PCB at a slight angle using a special silver epoxy that conducts current from the PCB pad to the electrodes. The device was designed to operate in six modes; digital high, digital low, 20, 1200 Hz, PWM 39% ON, PWM 94% ON. Each signals patterns generated by the device is shown in Fig. 4.5. The results showed that there are negative spikes at the falling edge due to stray inductance.

We conducted a preliminary study across fifteen participants to evaluate the effectiveness of digital taste as a means to stimulate taste. Participants chosen were of good health and reported no taste problems and were instructed not to eat, drink, or smoke two hours before the tests. As in Fig. 4.6, participant place the electrode to their tongue to receive the taste perception. We were interested in the robustness of such a system, such as sensory adaptation, controllability of the tastes through a population, and also the comfort of using this system.

Our study shows that some tastes can be effectively actuated across our population of users. Users showed some hesitance when placing the device into their mouths due to its appearance. Analytically and experimentally determine the characteristics of non-invasive electromagnetic stimulations on the human tongue towards electromagnetically generating primary taste sensations. At this first stage of the research, standard laboratory equipment for generation of electromagnetic fields were used, so

Fig. 4.3 Digital taste interface

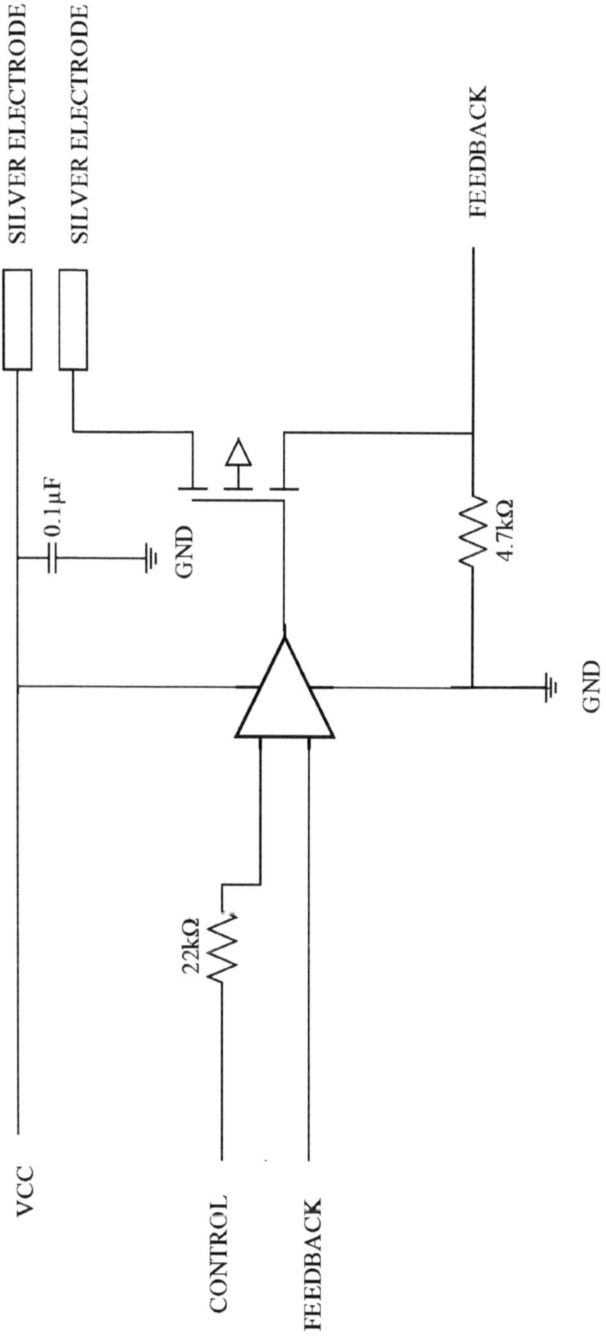

Fig. 4.4 Board diagram of the electric taste device

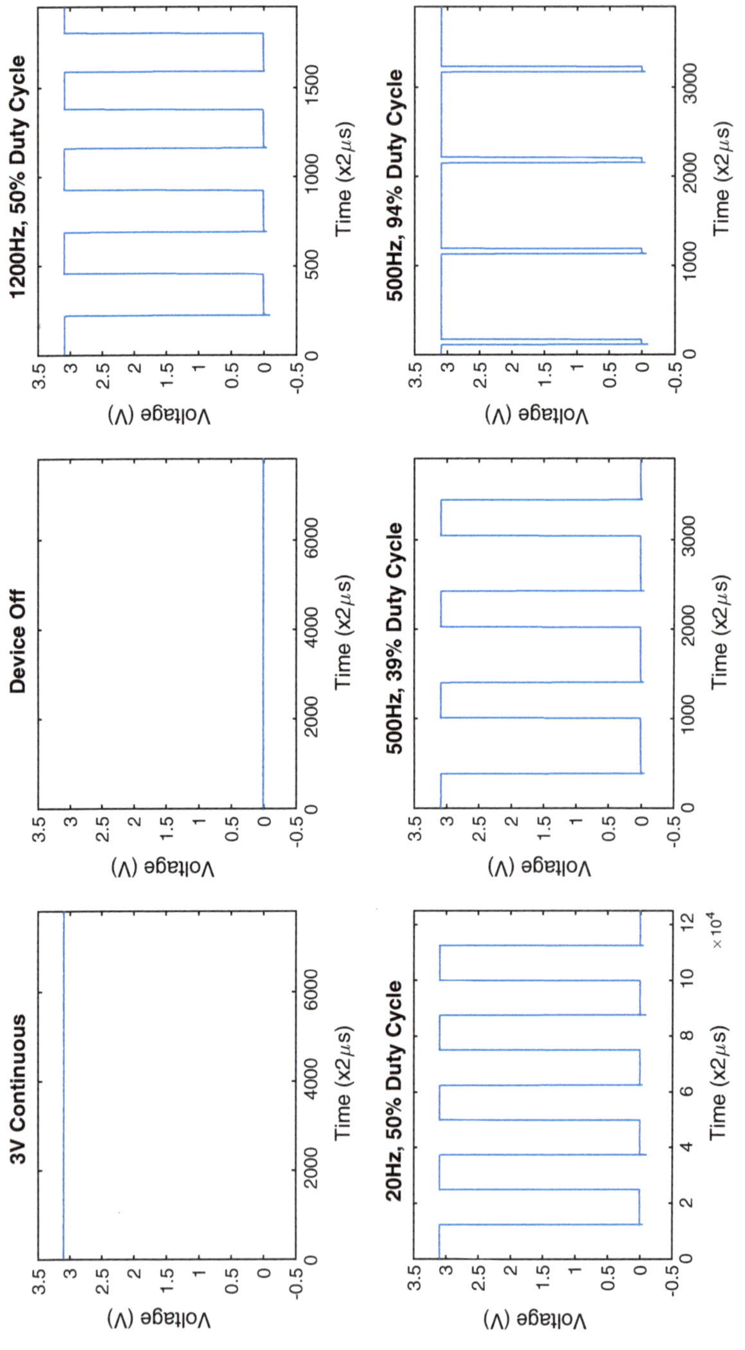

Fig. 4.5 Different stimulation signals generated by the device

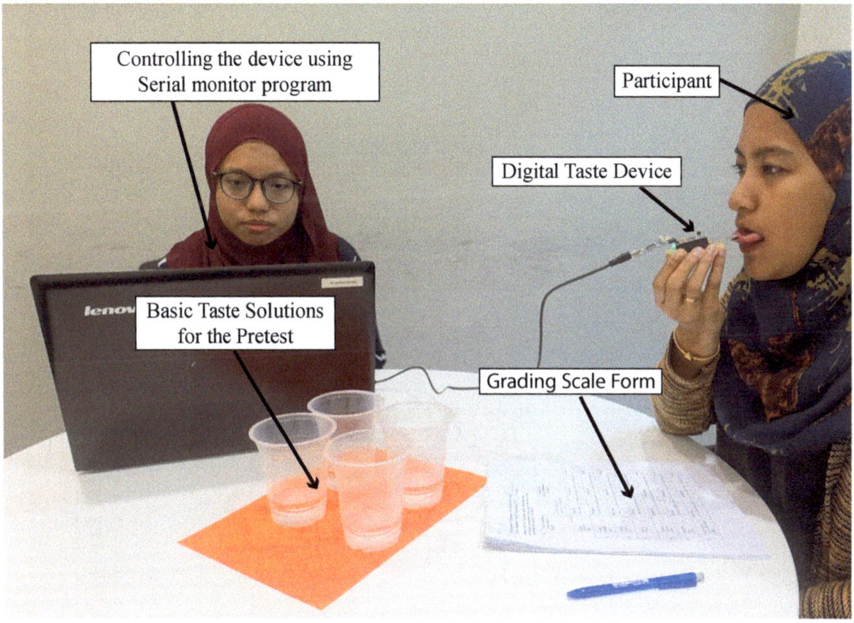

Fig. 4.6 User study setup of the electric taste experiment

the bulkiness and comfort will not be optimized. Safe and well established procedures were used throughout this research.

The percentages of the participants who reported the taste-related sensations and non-taste-related sensations for electrical stimulation is shown in Figs. 4.7 and 4.8. Additionally Figs. 4.9 and 4.10 report the means for the intensities of taste sensations and non-taste-related sensations produced by electrical stimulation. These results indicated that electrical stimulation of the tongue induced several taste sensations; 5 V stimulation actuate a strong metallic (86.67%) and electric (93.33%) taste sensations, 20 Hz stimulation evoked sour (86.67%), salty (83.33%), metallic (86.67%) and electric (93.33%) sensations, while 94% duty cycle stimulation induced sour (80%), metallic (86.67%) and electric (96.67%) sensations. Results showed that the effects were significant for sour($p < 0.001$), salty ($p < 0.001$), Metallic ($p < 0.005$). Other taste sensations reported included sweet, bitter, umami, fatty (oily), mint, carbonation, chemical, and spicy. Some participants experience non-taste perception using the same stimulation parameter such as coldness, numbing, unpleasant, and pressure.

In summary, we found that electrical stimulation had statistically significant effects on producing sour and salty sensations. In addition, more than 50% of subjects reported that electrical stimulation produce spicy, bitter, metallic, electric, and pressure sensations. We also found that increasing the voltage will produce more intense sensations. Further, lower duty cycles generated lingering and cold sensations while higher duty cycle values improved the pressure and unpleasantness.

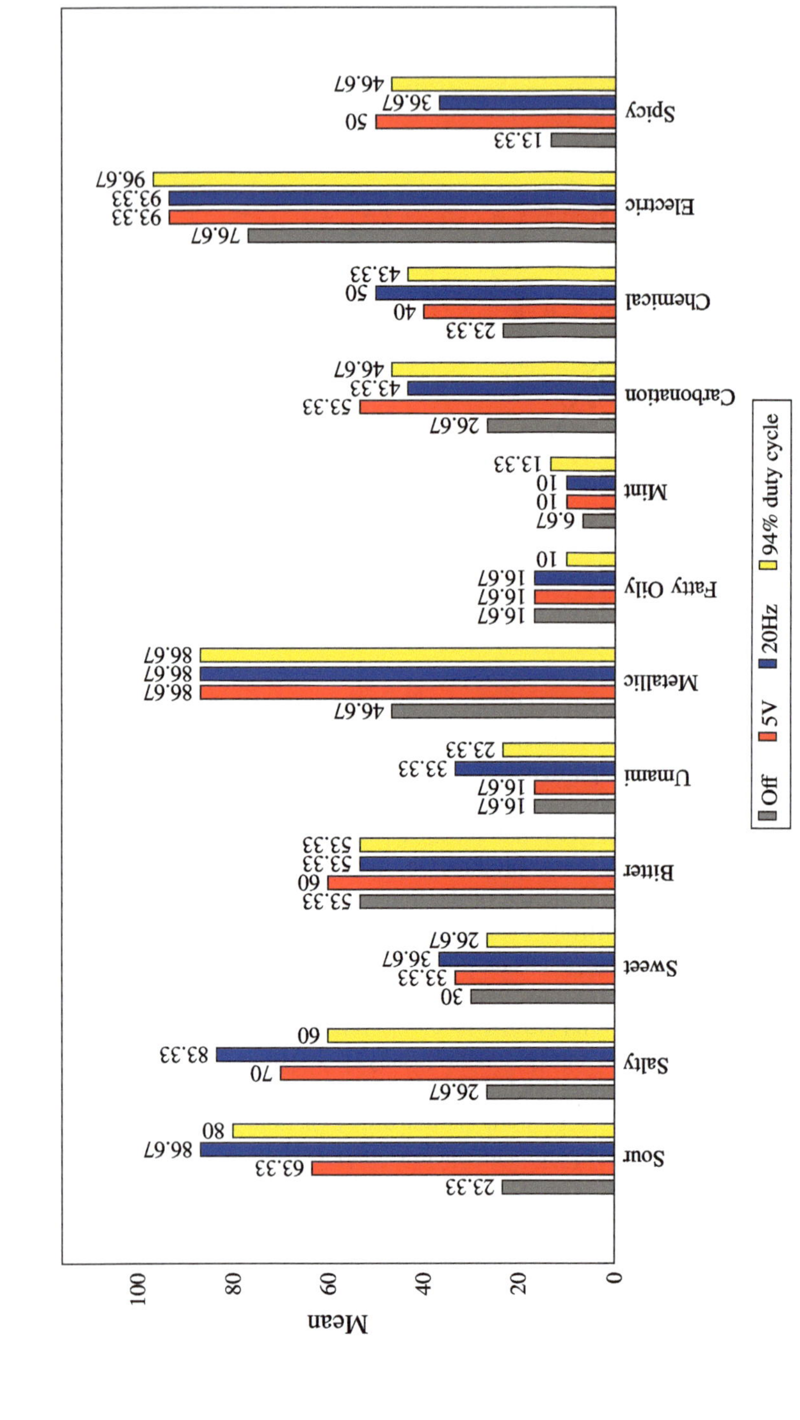

Fig. 4.7 Percentage of the taste related sensations reported for four different stimulations

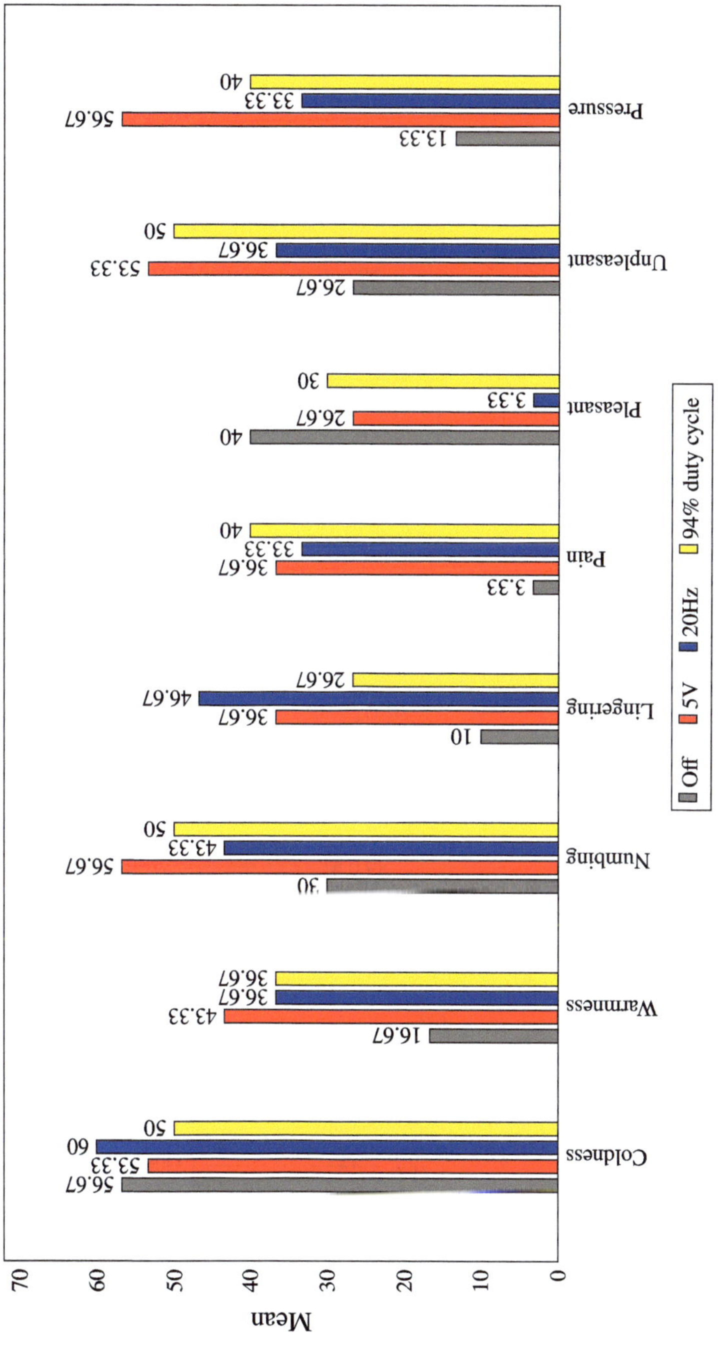

Fig. 4.8 Percentage of the non-taste related sensations reported for four different stimulations

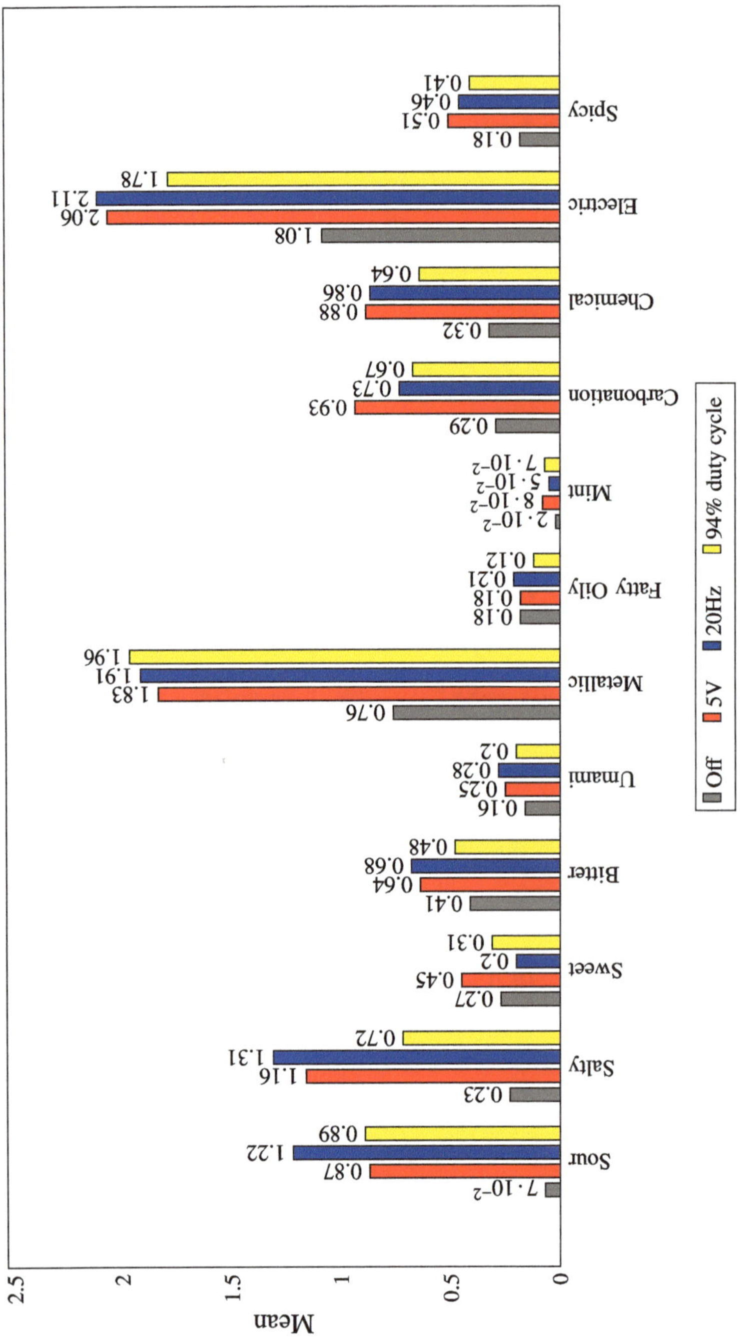

Fig. 4.9 Means of the taste related sensations reported for four different stimulations

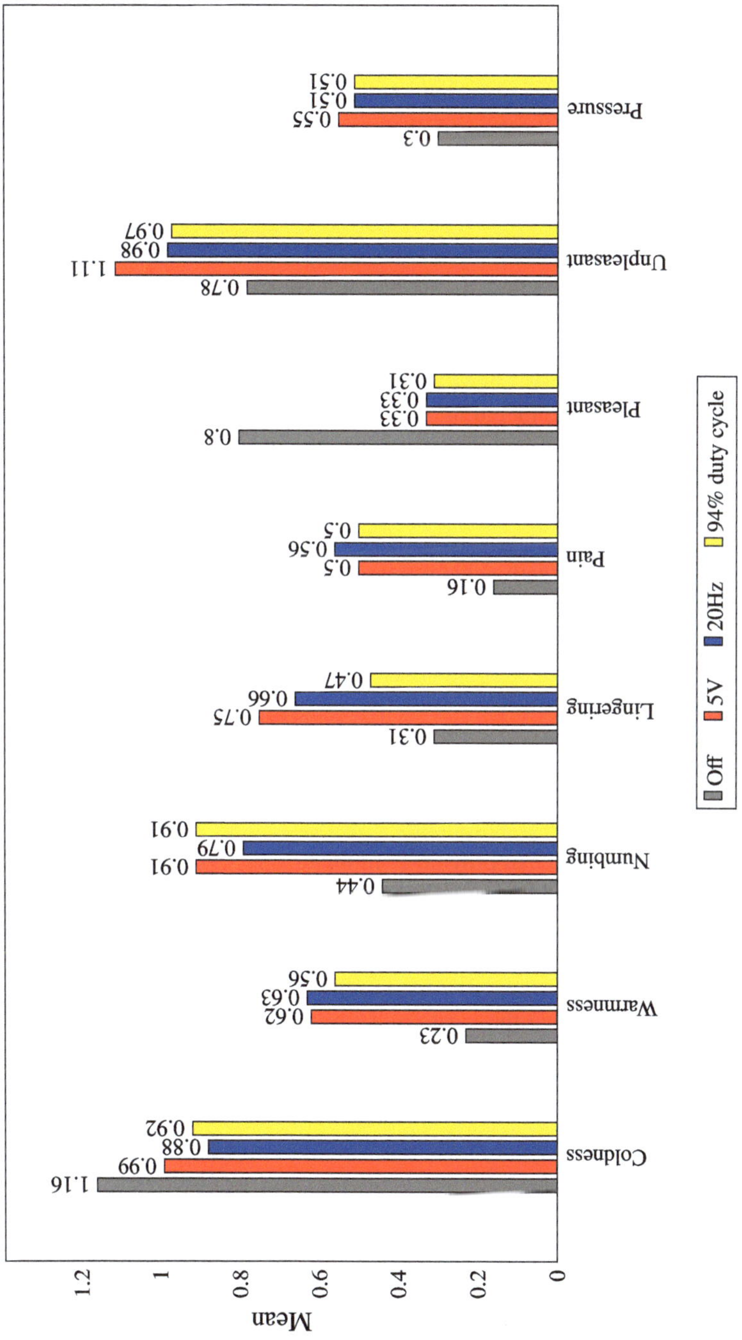

Fig. 4.10 Means of the non-taste related sensations reported for four different stimulations

The system was evaluated by comparing the perception of single tastes with real and artificial versions as a complete unit. In addition to user feedback to compare perception of real and artificial tastes, we will also examine the associated nucleus of the real and artificial tastes in the brain by Electroencephalography (EEG) [33], and functional Magnetic Resonance Imaging (fMRI) [34] techniques. In this way, we can also examine similarities and differences at the functional brain level of real and artificial tastes. Based on the EEG results, we will also examine latency and response of taste actuation input signals and perception in the brain.

4.4 Discussion

Electric taste technology can open up a multitude of new horizons and opportunities for research in the future, including in areas of human computer interfaces, entertainment systems, medical and wellness. Digital controllability of the sensation of taste provides a useful platform for engineers, interaction designers, and media artists towards developing multisensory interactions remotely, including the generation of new virtual tastes for entertainment systems. For scholarly research, this would help bring about answers to exactly what is language of taste.

We expect this research to culminate with a wearable unit that could clip inside the mouth. Users could wear it in daily lifestyle situations for augmentation. For example, Google Glasses promises a better experience of augmented reality unattainable with a smart phone. We can also realize accessories that would be attached to smart phones that could actuate a taste by placing the end of the apparatus in one's mouth. A friend could send you a taste over the internet by means of a social network and you could taste it electronically. They would simply input some sort of taste into the phone by text means or by selecting presets related to their current experience. Total controllable of the device can provide a stimulation of a taste from something that can't exist. This could also lead to a breakthrough in molecular gastronomy. Training molecular gastronomers is very difficult for most people, due to the requirement to be trained directly by experts at special facilities. Digital taste systems could do this by through training with a computer or through the internet. It could train ones sense of taste and possibly lead to the creation of new kinds of consumer edible delights.

We also believe patients who can't consume certain chemicals such as those with diabetes can stimulate the certain chemical in question (like sugar) taste in place of the actual chemicals. Our system could do so by augmenting preexisting tastes like seasonings without the danger of the actual chemical being present in the food. It may also be possible to cancel out certain tastes people wouldn't like, like bitterness, by adding more sugar to it, without worrying about consuming too much. This might allow humans to even consume certain foods that were previously unappetizing to eat or to encourage children to eat unpopular foods to maintain a healthy lifestyle.

As an input idea, people could experience flavors and the computer could output musical beats for both amusement and recording the notes to store our recipes in an interesting way. In this scenario, taste can be composed like a song. They can be composed to vary over time, for example, a sudden change of taste from salty to sweet, which cannot be achieved using real foods. People can make new kinds of recipes using purely digital compositions and post them on a social media site.

Users can create new knowledge using a digital taste machine and add pages embedded with the flavors similarly to adding pictures or sound to a webpage. People can rate the experiences by sharing them through another flavor. These experiences can gather to form more and more complex flavors and realize new potentials in gastronomy through our device. A chef or restaurant owner could introduce a new menu by sending alerts to fans of their work containing flavor information for new dishes. When the user opens the message, he or she can taste the new menu from their accessory as well as the food image.

By using taste stimulation we can create new ways to create content and new knowledge for food. We have conceptualized some ideas for how this may work in Fig. 4.11. Our ideas are Virtual SMS Menu, Culinary Education, Collaborative Remote Dining, Ambiguous Food, and approximating a service for future food printing at home.

Virtual Menu: Internet and mobile phone services can be integrated with our taste accessory to provide taste information for their digital contents. For example, when a restaurant introduces a new menu, they can conduct a special promotion using the taste accessory. They can send online coupons to their customers containing taste information for new dish and when the customer begins to read this message, he or she can taste the new menu from their accessory. Customers are able to check the restaurant menu before visiting through the taste accessory, by logging into the diners homepage. For a website this can be embedded into the site ads.

Culinary Education: Watching a recipe video gives us two sources of information visual and auditory. In the near future we could see audio visual systems that are expanded to utilize taste. You can check your own dish in progress by comparing the taste to the chef's dish. Similarly, we could add new dimensions to food entertainment

Fig. 4.11 Concept of digital taste in food application

media, when the lid comes off a steaming pot on your streaming video; you experience the same warm aroma and sensation.

Collaborative Remote Dining: We would be able to enhance remote co-dining and co-cooking experience, by translating tasting sense to music notes to record our daily recipe in an interesting way. We call this 'Taste symphony'. By capturing the taste from a curry cooked at one location and reconstructing and delivering in a personalize manner, in one scenario we envision a simple consumer technology that allows for these lonely individuals to cook meals collaboratively with their children remotely. By enhancing the taste generated in kitchen, decide up on a protocol to transmit taste and associate the ability of reconstructing the taste with the remote collaborative cooking space would recombine the creative nature of the humans all over the world and gives positive thinking of remote co-presence and co-living experience through this new kitchen space.

Digital Communication of Food: Sensing occurs between the senders environment and the media. The sensors can detect tastes from the environment or from specific foods. An example is that the various sensors in a kitchen can measure the tastes currently in the room. The sensed aromas and tastes can be communicated through internet after converting them into digital information and actuate on a specific user using the small wearable device. The 3D food printing could be the ultimate output, eventually. A chef could use the taste to give quick feedback of an ongoing recipe which when finalized could be sent to a 3D printing food. In the future we expect a home user to have such devices installed in their homes like with paper printer. A user could design his/her own food to be printed and send it to his/her friends printing device. While constructing the food to save resource they can approximate the food with taste device. They could even upload this as a work in progress to allow other to quickly sample the food for approval.

4.5 Conclusion

Audio, visual, and haptic fields are well researched. We believe the next challenge is virtual taste using techniques that are currently unattainable using chemical based methods. Digital taste is one such technique of actuating taste with electric stimulation. This has advantages over chemical stimulation because it is more manageable and taste can be transmitted directly over the internet. We believe through this research we can eventually digitally actuate all main tastes (bitterness, saltiness, sourness, sweetness, umami) using digital means to actuate and transmit the taste information. It could have long term implications in the future of many disciplines. By using digital taste, we will be able to create new ways to create content and new knowledge of food that can be shared, learned, and created in new interesting ways previously unobtainable with current methods.

References

1. Bottino J Television as a media technology final paper
2. Murray MA (2017) Our chemical senses: taste
3. Price S, Desimone JA (1977) Models of taste receptor cell stimulation. Chem Senses 2(4):427–456
4. Ellegård EK, Goldsmith D, Hay KD, Morton RP (2007) Studies on the relationship between electrogustometry and sour taste perception. Auris Nasus Larynx 34(4):477–480
5. Frings S (2010) The sour taste of a proton current. Proc Natl Acad Sci 107(51):21955–21956
6. Salata J, Raj J, Doty R (1991) Differential sensitivity of tongue areas and palate to electrical stimulation: a suprathreshold cross-modal matching study. Chem Senses 16(5):483–489
7. Lindemann B (1996) Taste reception. Physiol Rev 76(3):719–766
8. Levy LM, Henkin RI, Hutter A, Lin CS, Schellinger D (1998) Mapping brain activation to odorants in patients with smell loss by functional mri. J Comput Assist Tomogr 22(1):96–103
9. Sakurai K, Aoyama K, Mizukami M, Maeda T, Ando H (2016) Saltiness and umami suppression by cathodal electrical stimulation. In: Proceedings of the 1st workshop on multi-sensorial approaches to human-food interaction. ACM, p 2
10. Von Békésy G (1966) Taste theories and the chemical stimulation of single papillae. J Appl Physiol 21(1):1–9
11. Lawless HT, Stevens DA, Chapman KW, Kurtz A (2005) Metallic taste from electrical and chemical stimulation. Chem Senses 30(3):185–194
12. Kortum P (2008) HCI beyond the GUI: design for haptic, speech, olfactory, and other nontraditional interfaces. Morgan Kaufmann
13. Ranasinghe N, Cheok A, Nakatsu R, Do EYL (2013) Simulating the sensation of taste for immersive experiences. In: Proceedings of the 2013 ACM international workshop on Immersive media experiences. ACM, pp 29–34
14. Maynes-Aminzade D (2005) Edible bits: seamless interfaces between people, data and food. In: Conference on human factors in computing systems (CHI'05)-extended abstracts, 2207–2210
15. Derbyshire D (2009) Revealed: the headset that will mimic all five senses and make the virtual world as convincing as real life. Daily Mail:03–05
16. Comeaux D, Ross J The virtual reality bim immersion experience
17. Narumi T, Sato M, Tanikawa T, Hirose M (2010) Evaluating cross-sensory perception of superimposing virtual color onto real drink: toward realization of pseudo-gustatory displays. In: Proceedings of the 1st augmented human international conference. ACM, p 18
18. Narumi T, Kajinami T, Tanikawa T, Hirose M (2010) Meta cookie. In: ACM SIGGRAPH 2010 Posters. ACM, p 143
19. Nugent PM (2013) Gustometer in psychologydictionary.org
20. Spector AC, Blonde GD, Henderson RP, Treesukosol Y, Hendrick P, Newsome R, Fletcher FH, Tang T, Donaldson JA (2015) A new gustometer for taste testing in rodents. Chem Senses 40(3):187–196
21. Volta A (1800) On the electricity excited by the mere contact of conducting substances of different kinds. in a letter from Mr. Alexander volta, FRS Professor of Natural Philosophy in the University of Pavia, to the Rt. Hon. Sir Joseph Banks, Bart. KBPRS. Philos Trans R Soc London:403–431
22. WIKI (2006) Manuel: alessandro giuseppe antonio anastasio volta https://commons.wikimedia.org/wiki/File:Alessandro_Volta.jpeg. Accessed 1 Feb 2018
23. Plattig KH, Innitzer J (1976) Taste qualities elicited by electric stimulation of single human tongue papillae. Pflügers Archiv 361(2):115–120
24. Nakamura H, Miyashita H (2011) Augmented gustation using electricity. In: Proceedings of the 2nd augmented human international conference, ACM, p 34
25. Nakamura H, Miyashita H (2013) Enhancing saltiness with cathodal current. In: CHI'13 extended abstracts on human factors in computing systems, ACM, pp 3111–3114

26. Nakamura H, Miyashita H (2013) Controlling saltiness without salt: evaluation of taste change by applying and releasing cathodal current. In: Proceedings of the 5th international workshop on Multimedia for cooking & eating activities, ACM, pp 9–14
27. Aruga Y, Koike T (2015) Taste change of soup by the recreating of sourness and saltiness using the electrical stimulation. In: Proceedings of the 6th augmented human international conference, ACM, pp 191–192
28. Tomita H, Ikeda M, Okuda Y (1986) Basis and practice of clinical taste examinations. Auris Nasus Larynx 13:S1–S15
29. Burgmans T (2005) Method and apparatus for simulating taste sensations in a taste simulation system. US Patent App 10/597,422, 26 Jan 2005
30. Huo X, Wang J, Ghovanloo M (2007) A wireless tongue-computer interface using stereo differential magnetic field measurement. In: 29th annual international conference of the ieee engineering in medicine and biology society, EMBS 2007, IEEE, pp 5723–5726
31. Kim D, Tyler ME, Beebe DJ (2005) Development of a tongue-operated switch array as an alternative input device. Int J Hum Comput Inter 18(1):19–38
32. Ranasinghe N (2013) Digital lollipop
33. Klemm WR, Lutes SD, Hendrix DV, Warrenburg S (1992) Topographical EEG maps of human responses to odors. Chem senses, Oxford University Press, 17(3):347–361
34. Levy LM, Henkin RI, Lin CS, Hutter A, Schellinger D (1998) Increased brain activation in response to odors in patients with hyposmia after theophylline treatment demonstrated by fMRI. J Comput Assist Tomogr, LWW, 22(5):760–770

Chapter 5
Thermal Taste Interface

Abstract This chapter presents a new taste device for digital taste communication called 'Thermal Taste Interface'. It produces thermal taste sensations on the tongue purely by modifying the temperature of the surface of the tongue (from 25 to 40 °C while heating and 25 to 10 °C while cooling) within a short period of time. Our results suggested that rapidly heating the tongue produces sweetness, fatty/oiliness, electric taste, warmness, and reduces the sensibility for metallic taste. Similarly, cooling the tongue produced mint taste, pleasantness, and coldness. By conducting another study on the perceived sweetness for sucrose solutions after the thermal stimulation, we found that heating the tongue significantly enhances the intensity of sweetness for both thermal tasters and non-thermal tasters. Also, we found that faster temperature rises on the tongue produce more intense sweet sensations for thermal tasters. This device offers easy customization options such as rapid heating and cooling, different stimulation speeds, extended temperature range (from 4 to 100 °C), and ability to integrate and control using a software. The sections below discuss the development, technical evaluation, user evaluations, and future work of this device. We believe this technology may can the user experiences related to thermal taste in different disciplines including Human-Computer Interaction, New Media Arts, Communication and Medicine.

5.1 Introduction

Sweet taste has been identified as the most pleasant out of all the tastes [1]. Therefore, the stimulation of sweet taste is a crucial element in digital communication. Over the years, there were some attempts made by researchers to digitally produce or modify taste sensations [2–4]. However, a pure digital device that can repeatedly stimulate sweet taste without using chemicals has not been developed. Therefore, researchers used chemical based interfaces that stimulate sweetness in HCI [5–7]. However, some recent works have identified that thermal stimulation can evoke sweet sensations on the tongue. This is achieved by triggering the TRPM 5 channel (Transient receptor potential cation channel subfamily M member 5). In some previous experiments in

A. D. Cheok and K. Karunanayaka, *Virtual Taste and Smell Technologies for Multisensory Internet and Virtual Reality*, Human-Computer Interaction Series,
https://doi.org/10.1007/978-3-319-73864-2_5

Fig. 5.1 Concept Image for our objective: In future, people would be able to communicate and experience the sweetness digitally

the medical field suggests that the increasing temperature on the tongue over the time has reported sweet sensations [8], and higher temperatures resulted enhancing the sensitivity for sweetness [9, 10].

Today internet communication is based on text, audio, and visual data. Digitizing the taste sensations will allow us to transmit and regenerate taste sensation over digital networks. Chemical taste actuation interfaces has drawbacks such as refilling, precious controlling of chemicals, provide unnecessary calories, chemicals or minerals to the body. These limitations were motivated us to develop digital interfaces for taste.

To address the above mentioned issue and limitations we developed a device that can induce thermal taste sensations. This interface stimulate the surface of the tongue by the means of temperature. We increased the temperature of the silver plate from 25 to 40 °C during heating and reduce from 25 to 10 °C while cooling. By placing this device on top of the tongue, it results a slight modification of the temperature on the tongue surface and we have observed that 'Thermal Tasters' reported mainly sweetness, minty and fatty tastes. Further, participants also reported the enhancement of sweetness. Faster the temperature rise resulted more intense sweet taste sensations. We believe by continuously improving this technology we may be able to create, enhance, and share sweet taste experiences with others remotely as illustrated in Fig. 5.1.

5.2 Related Works

Cruz and Green [8] are the first researchers who showed the sweetness can be invoked as a result of thermal stimulation. They have studied the effects of stimulation the human tongue using temperature range from 15 to 35 °C with approximately 1.5 °C/s speed and the subjects reported sweet taste while increasing of the temperature.

This experience has explained as 'phantom taste' of sweetness. Also the people who experienced this kind of taste were started to refer as 'Thermal Tasters'.

Talavera et al. [9] showed that the increasing temperature has resulted activation of TRPM5 channel that generates a depolarizing potential in the taste receptor cells. This effects causes the enhanced sweetness perception at high temperatures and 'thermal taste', the phenomenon whereby heating or cooling of the tongue evoke sensations of taste in the absence of chemical tastants. It has been concluded that stimulating TRPM5 with temperature as an input results different taste sensations as outputs. It has been further suggested that other tastes, such as salty and sour, perceived by TTs, may be linked to the temperature sensitivity associated with the of the channels involved in their chemical transduction. The cold and heat activation of the receptors can be the fundamental of the temperature dependence of salt responses. Enhancement of sweet perception by using warm temperatures cause the heat to activate TRPM5 channel [11]. Applying thermal sensations to the skin around nose resulted with charming or unpleasant feelings. The 'Affecting Tumbler' [4] is an example for that.

Even though the thermal effect with TRPM5 channel has been studied in medicine, a proper controlled computer system that can reproduce thermal taste sensations has yet to be developed. Having said that, authors [12] experimented with electrical and thermal stimulation in combination on the tongue, and showed that it is possible to generate four of the basic tastes: sour, sweet, bitter, and salty. During the experiments conducted in [13], stimulation of the tongue with heating and cooling the tongue, participants reported sensations of sweetness and sourness. The first sweet-specific device using thermal stimulation was proposed by the authors of this book in 2015 [14].

Our approach is different from the works mentioned above in many ways. Our main objective is to develop a controllable and repeatable digital technology to generate taste sensations. We decided that this technology should be a device that can be plugged into computers and it should be able to be programmed and controlled through the computer. Therefore, our work is different from the previous works from the medical field. Also the works from HCI are only concentrated on inducing the sweet taste sensations. In this research we consider Thermal Taste as a combination of taste and non-taste related sensations. Therefore, one of our objective is to describe the thermal taste using known taste and non-taste sensations. Further, in this research we study how thermal stimulation can change the intensity of sweetness and also how different speeds of temperature rising can affects the sweetness. In addition, our final objective is to integrate this thermal taste technology for VR and multisensory communication. According to our knowledge there was no attempt been made so far to achieve this.

5.3 Method

5.3.1 Hardware Module

The basic idea of this interface was to design a device that can heat up the surface of the tongue and stimulate the TRPM5 taste receptors within short period of time. The latest prototype of the thermal sweet taste interface is shown in the Fig. 5.2. Thermal Taste interface consists of an electronic controller circuit, silver plate that connects to the Peltier, liquid cooling fan and a software module.

As shown in Fig. 5.2, our proposed device consists of an Arduino Pro-Mini microcontroller, an FTDI serial interface to communicate with the PC, a silver plate that is attached to a Peltier module, a liquid cooler system, a high current DC motor driver and a temperature sensor.

The temperature on the tongue is changed by adjusting temperature of Peltier module that are tightly coupled with a silver electrode as shown in Fig. 5.3. When the electric current passes across a junction between two materials in the peltier module, one side of the device is emitting the heat while other side absorbs. In this experiment, we used the heat emitting side of the peltier to raise the temperature of the peltier and eventually the tongue. By alternating the direction of the current, it is possible to change the heating and cooling sides of a peltier. We use this option to rapidly cool down and reset the device to the original state. The peltier device we have

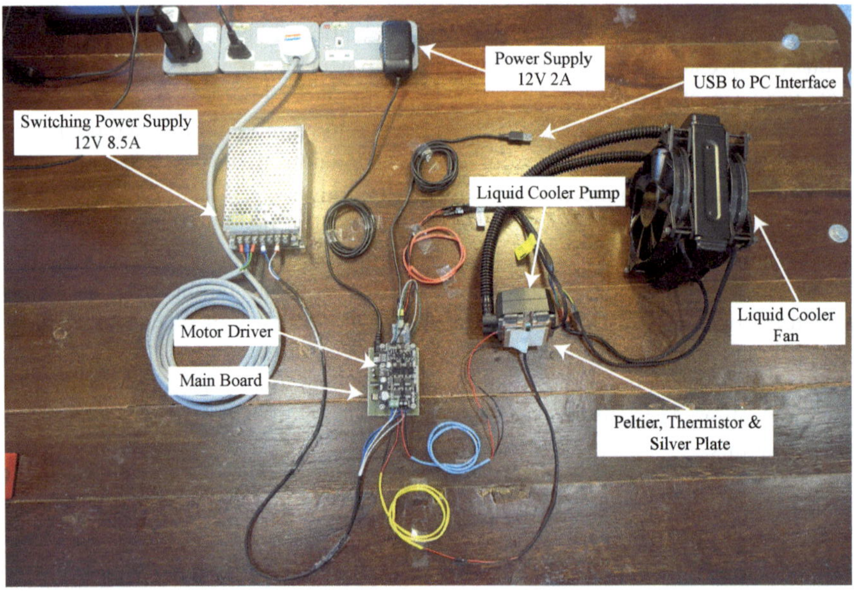

Fig. 5.2 The latest prototype of the thermal taste interface device

selected can reach the temperatures from −40 to 80 °C while consuming current up to 8.5 A with 15.6 V for 100% duty cycle. Therefore, a high current DC motor driver has been used to drive the Peltier. This motor driver allowed bi-directional controlled of its motor and the speed control PWM frequency up to 20KHz. The direction of the motor made it possible to control heating and cooling part of the peltier. PWM input from Arduino microcontroller programming can control the duty cycle and current of the peltier.

The transferring of the heat generated by the peltier module to the tongue was done by using a silver plate. This plate was mounted on top of the heating side of the peltier device as shown in Fig. 5.3. We have used Silver due to two reasons; no metallic sensations and less specific heat. Unlike other metals such as copper, gold or stainless steel, silver does not produce high metallic sensation to the tongue. This provides us an advantage that when people reports our device they report they actually felt as a result of the stimulation and not the metallic sensation usually you

Fig. 5.3 The silver plate is assembled together with peltier device and liquid cooler pump

felt with other metals. Also silver has less specific heat so it can rise the temperature fast and efficiently pass the heat to the tongue. Previously we have used copper for our earlier prototypes but when compared with copper, silver has less specific heat (he specific heat of copper is about 0.092 kcal/kg °C while silver is 0.057 kcal/kg °C [15]). Also, we made the plates very thin (0.5 mm) and the area of the stimulation smaller 2 cm by 4 cm to make the plate less in weight so we need less energy to heat up the silver.

The heating up and cooling down of the device is controlled by the microcontroller firmware by monitoring the temperature of the temperature sensor that attached to the silver plate. Once again we have used thermal epoxy to attach the sensor with the silver. This is to obtain the exact temperature of the silver so that we did not burn users tongue. We make sure the device is prevented from over heating and over cooling. PID controlling [16] technique has used to control the current provided to Peltier over the time. To make the heating process more efficient a liquid cooler was mounted to the cooling side of the peltier module. This improves absorbing the cool generated from the cooling side which indirectly improves the heating of the other side.

There are three different power levels has been used for the circuit. All the logic signals and power to the microcontroller was USB powered and it uses 5 V (with less than 500 mA). To operate peltier, switching power supply of 12 V and 8.5 A is used. For liquid cooling, DC power supply of 12 V and 2 A was used. Therefore, to decouple the logic and the power we used optocouplers in the circuit. Optocouplers prevent DC current from flowing to the other components when it is short circuited as detailed in Fig. 5.4.

5.3.2 Software Module

5.3.2.1 Device Firmware

The firmware that runs on the microcontroller was implemented using Arduino. Basically the Arduino program receives signals from the computer for heating up or cooling down and then the program starts to activate the Peltier module and monitors using PID controlling the temperature sensor input. ‘Putty’, ‘Hyperterminal’, ‘Chrome Serial’ or any serial port interfacing client can be used to control the device. The firmware allows users to easily customize the functionalities of the device like change heating, cooling, threshold settings, stimulation intervals, and program stimulation protocols.

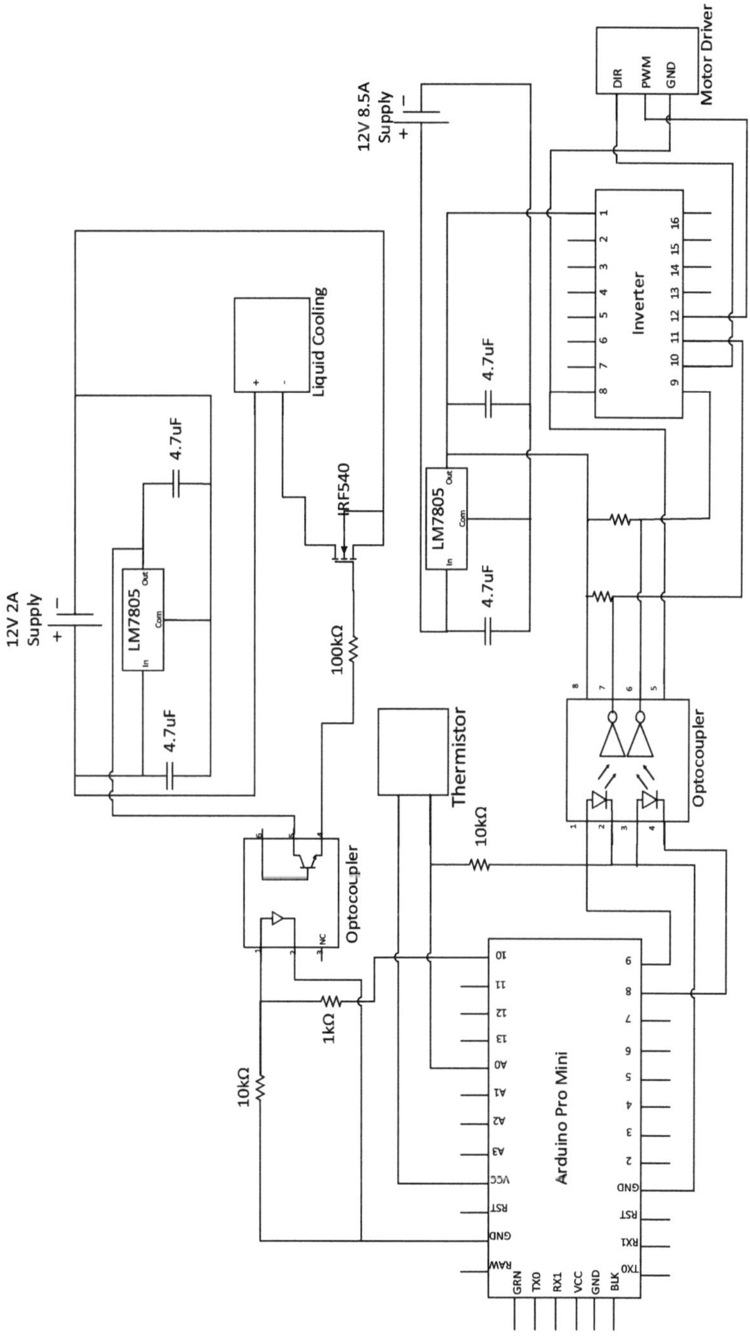

Fig. 5.4 The schematic drawing of the thermal taste interface

5.4 Technical Evaluation of the Device

We have conducted five technical experiments during the development of this device. Those studies include exploring the minimum and maximum temperature of the device, finding suitable PID parameters for heating and cooling, the time difference for different starting temperatures, different stimulation speeds using different PID control parameters and changing the temperature of solutions using different plate designs. The following sections will discuss about these experiments in detail.

5.4.1 Finding the Temperature Limits

This experiment studied the highest and lowest temperature that can be stimulated by the device. The device was operated with maximum PWM value. The results of this experiment is illustrated in the Fig. 5.5. The maximum temperature obtained during the study is 100 °C and the lowest was 4 °C. This experiment could not achieve the limits defined by the Peltier manufacturer (−40 and 80 °C). The main reasons for this is we are not running the Peltier module with maximum voltage. The Peltier can provide up to 15.6 V our device was consuming only 12 V. So the number of watts consumed by the device was 77% of the maximum possible consumption. This is mainly because we mounted the silver plate on the original heating side of the Peltier, so the cooling will not be as efficient compared with using the original cooling side of the Peltier. Further, limitations of the heat sink, liquid cooler design, and efficiency of the thermal epoxy could be the other reasons for this.

5.4.2 Finding the Best PID Control Parameters for the Device

A PID controller is a control loop feedback mechanism where it continuously calculates an error value as the difference between a desired setpoint and a measured process variable [16]. This experiment was designed to identify the best PID controller parameters for the device.

During this experiment, we were trying to achieve two aims; faster response time and better stability. However, in practical PID control systems this is impossible to achieve both the goals simultaneously. Therefore, we had to find parameter which provides acceptable stability with better speed of temperature rise.

Each of the PID parameters time and temperature from table above are analysed and calculated to get the shortest rise time. The graphs are plotted using the data obtained from experiments. The calculation are based on the 10 and 90% of temp difference to find the rise time.

PID parameter of Kp, Ki and Kd are manually tuned to find the shortest rise time and faster settling time. For heating process, Kp is increased until the temperature

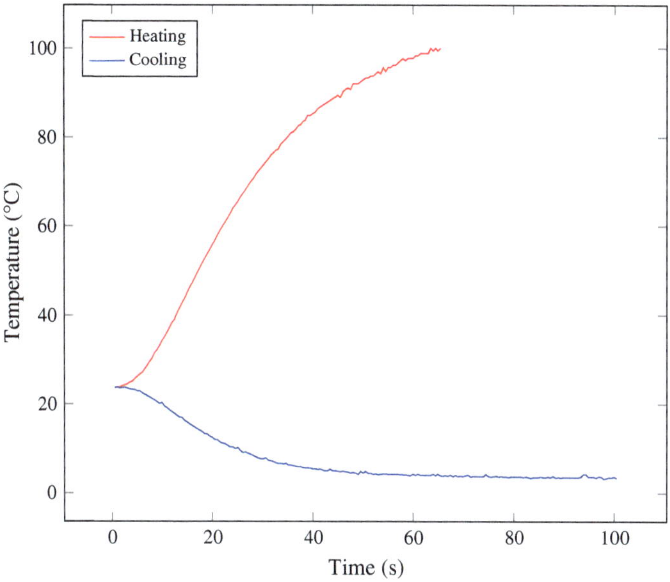

Fig. 5.5 The highest and lowest temperatures that the device can achieve

start to oscillate from 25 to 40 °C. Then, Ki is increased to correct the offset in the given time for the system. The heating process experiments were started when the temperature for Peltier and silver electrode is around 25 °C. And the system is left running for few minutes until it reached stable 40 °C. The first oscillation of 25 to 40 °C were tabulated to calculate the rise time.

The experiments for cooling process are started after heating process. As for cooling process, the step to find parameter for Kp, Ki and Kd are similar. It is the reverse operation of heating process. Since the setpoint temperature of heating process is 40 °C, the data were tabulated started from 40 °C until it reached stable 25 °C. The first oscillation of 40 to 25 °C are collected to calculate the fall time of the system.

By experimentally testing for different PID combinations, we concluded that the most suitable PID parameters for heating were Kp = 20, Ki = 10, and Kd = 30 with a 6.0 s rise time (set temperature of 40 °C), and the most suitable PID parameters for setting back the temperature to 25 °C after heating were: Kp = 60, Ki = 0, and Kd = 50 (set temperature of 25 °C). Further, we found that the most suitable PID parameters for cooling as Kp = 1000, Ki = 300, and Kd = 8000 (set temperature of 10 °C), and for setting back the temperature back to 25 °C after cooling as: Kp = 20, Ki = 10, and Kd = 30 (set temperature 25 °C). Figure 5.6 shows how temperature changes with the time for each set of PID parameters while reaching the desired temperatures (40, 25, and 10 °C). During our experiments we used the rising and falling sections of the temperature curves (referring to the blue and red color curves in Fig. 5.6) for stimulating the subjects. We used the orange and cyan color curves

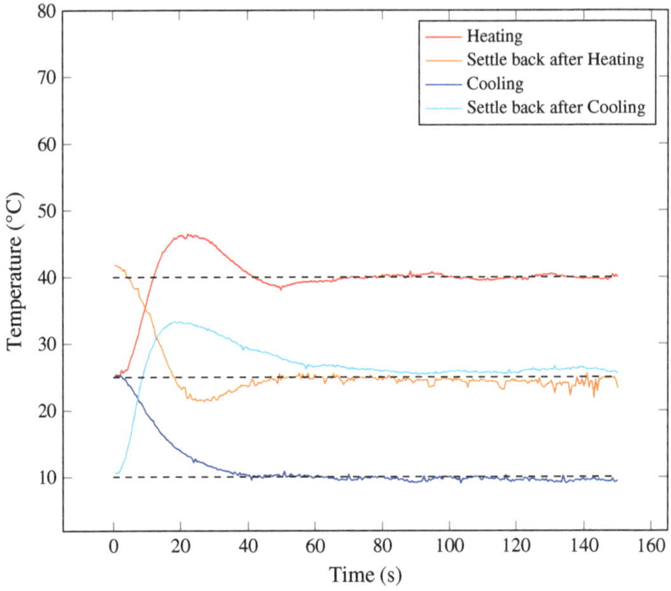

Fig. 5.6 Very suitable PID control curves obtained for this device for heating, cooling, and settling back the device to 25 °C after heating and cooling

to set the device back to 25 °C. These selected PID parameter sets resulted in 5% steady state error of the setpoint, less overshoot, and a very quick settling time.

5.4.3 Study How the Starting Room Temperature May Affect with PID Controlling

We have noticed that based on the starting temperature of the device may take different time to settle at 40 °C. Therefore, we have decided to study the difference in time it takes for temperature rising with three different room temperatures. Those temperatures are at 26.5, 21 and 15.2 °C. We have used the same PID control values obtained by the previous experiment. The results we have obtained is shown in Fig. 5.7. It shows that 21 °C curve shows a less rise time and probably best suited for this kind of experiment because it provides comparatively higher Celsius per second change. Therefore, we have decided to set a lower setpoint for the device at 20 °C.

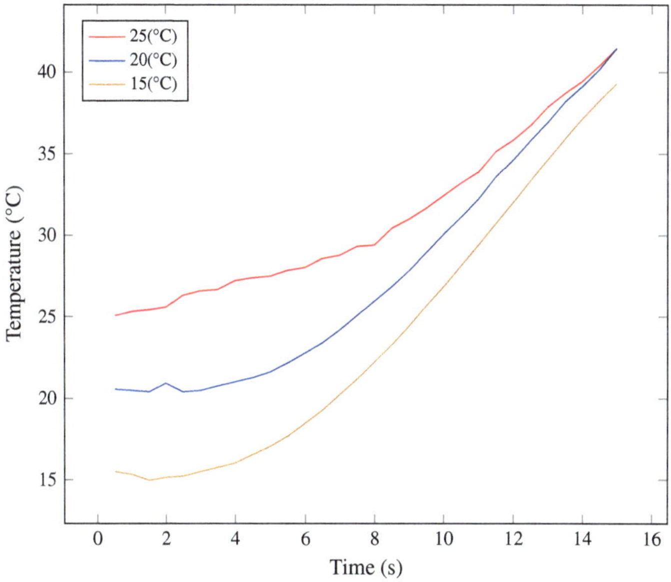

Fig. 5.7 This graph shows how different starting temperatures can affect the rate of the temperature change

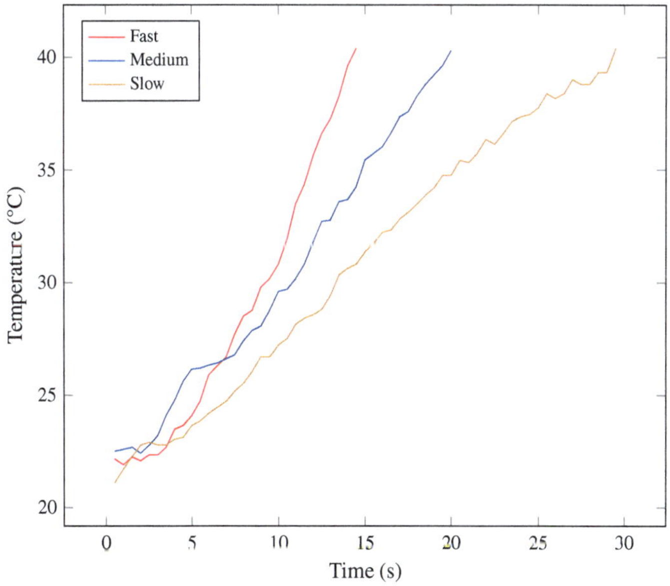

Fig. 5.8 Three different stimulation speeds used for the final study

5.4.4 *Finding Three Different Stimulation Speeds for the Third User Study*

This experiment has done for the obtaining three different stimulation speeds for the device, so we kept the same PID parameters we selected in the previous experiment but we have modified the current provided to the peltier module by limiting the final output PWM to the to 60 and 40%. The temperature rise curves of three different speeds are shown in the Fig. 5.8. Here the 'fast' curve is based on the PID parameters Kp = 20, Ki = 10, Kd = 30 with 6.0 s rise time. The 'medium' curve with output PWM to 60% has resulted 10 s rise time. The 'low' curve with PWM to 40% resulted 15 s rise time.

5.5 User Evaluation

We conducted three different user evaluations for the thermal taste device. First, we did a characterization of the thermal taste sensations produced by this device. We have identified people who can receive sweet sensations purely from the Thermal Taste Machine. Second, we studied whether the thermal taste device can enhance the intensity of sweet taste sensations produced by the sucrose. Third experiment, we tested participants to rate the sweet sensations they perceive for different speeds of temperature change. All the experiments were conducted according to guidelines set by the Imagineering Institute's ethics board.

5.5.1 *Characterization of the Thermal Taste*

For this experiment, we used Thermal Taste Interface to stimulate the surface of the tongue using three different ways. First was the heating, we increased the temperature from 25 to 40 °C. Second was the cooling, we decreased the temperature from 25 to 10 °C. Third was the device 'OFF' state in which the silver plate was kept at room temperature during the stimulation. Total of 39 participants consist of 16 males and 23 females were participated in this user study. Majority of them were recruited from Universiti Teknologi Malaysia (UTM) who were between 20 to 23 years old ($M = 24 \pm 6.9$).

5.5.1.1 Apparatus and materials

A pilot study was conducted prior to the user study to make sure we were familiar with the user study steps and protocol. This pilot study was conducted with 10 naive subjects with similar arrangements as the user study. The stimulation order was randomized between three parameters and each participant undergoes 12 trials. There were 4 trials for each stimulation condition.

Every participants completed a questionnaire which requires them to try four different taste solutions (sweet, sour, salty and bitter solutions) and identify the taste. This pre-screening was done to exclude any subject who were having taste dysfunction before proceed to the user study. Participants were tasted each solutions one after another and identified the taste. In between each trial, they needed to rinse their mouth with distilled water to remove the residue of the solutions on the tongue. All 39 participants were cleared to take part in the user study after the pre screening.

The user study was done in closed room where the participants were made to seat comfortably on a chair. Then, they were asked to rinsed their mouth prior to the first stimulation. They placed the silver electrode on their tongue and the experimenter activated the device with the first stimulation. Then, the participants were asked to rate the sensations they perceived on the intensity recording sheet in which it had 20 different sensations on the scale. This procedure was repeated for the next eleven trials and prior every trials, participants rinsed their mouth with distilled water.

The percentages of the participants who reported the taste-related sensations and non-taste-related sensations for thermal stimulation is shown in Figs. 5.9 and 5.10. The means for the intensities of taste sensations and non-taste-related sensations produced by thermal stimulation is shown in Figs. 5.11 and 5.12. One-way ANOVA was used to compare the means between three different stimulations: device switched off, heating, and cooling. Results showed that the effects were significant for sweetness ($p = 0.003$), fatty-oiliness ($p < 0.001$), minty ($p = 0.005$), electric ($p = 0.04$), heating ($p < 0.001$), cooling ($p < 0.001$), and pleasantness ($p = 0.005$). These results indicate that heating of the tongue induced sweet, fatty (or oily), electric, and warm sensations, while cooling of the tongue resulted in minty, cooling, and pleasant sensations. A previous experiment with mice suggested that the long-chain unsaturated free fatty acid and linoleic acid (LA) could depolarize the TRPM5 channel [17]. Also, it showed that mice lacking TRPM5 exhibit less sensitivity for LA. Therefore, we can argue that similar fatty taste sensations were produced in the human subjects by thermal stimulation. Further, heating the tongue showed positive results for producing spicy, numbing, and unpleasant sensations. In addition to that, heating the tongue reduced sensibility for metallic taste compared with cooling or off states. A similar result was discussed earlier for the thermal tasters in [18]. By using different types of heating, these effects can be further explored and enhanced in the future.

5.5.2 *Enhancement of Sweet Taste Using Thermal Stimulation*

Previous experiments from the medical field reported that thermal stimulation can enhance sweet taste sensations [9, 19]. The reason for this is that the TRPM5 cation channel shifts the midpoint for voltage-dependent activity to negative voltages when

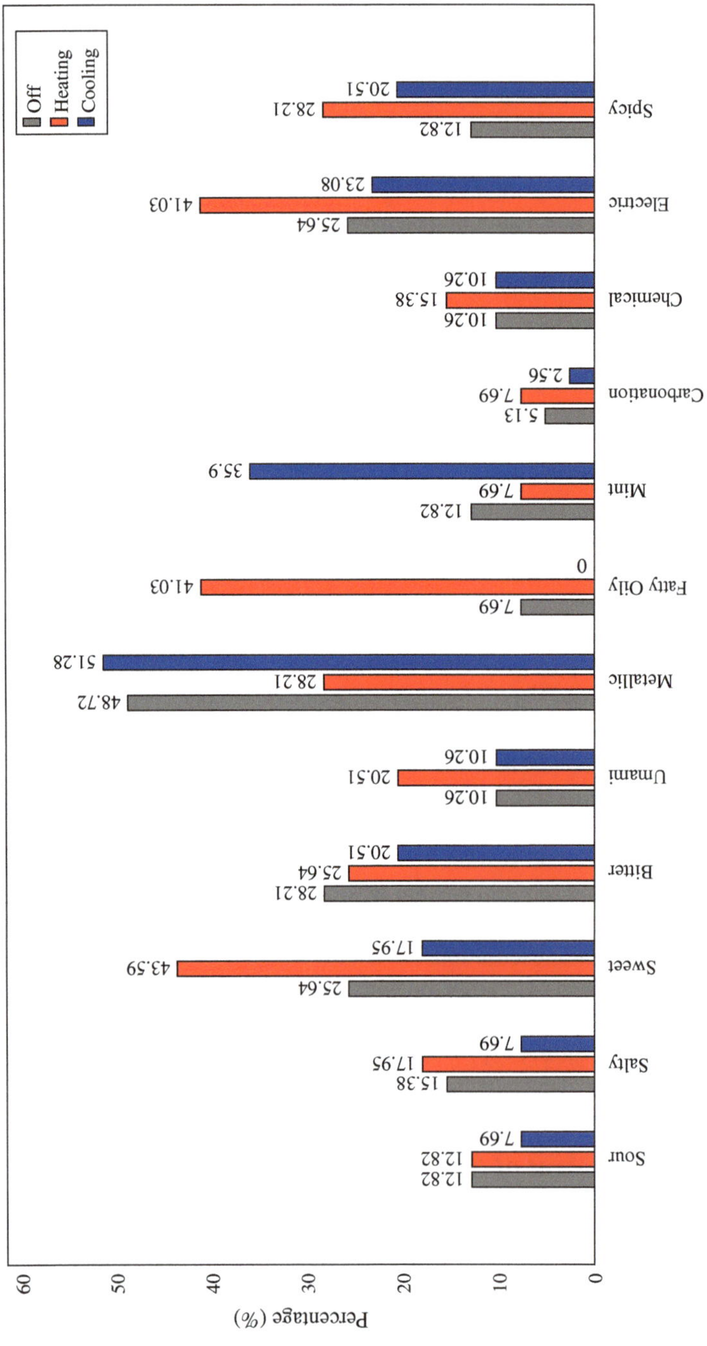

Fig. 5.9 Percentage of participants who reported taste related sensations for three different stimulations; off, heating, and cooling

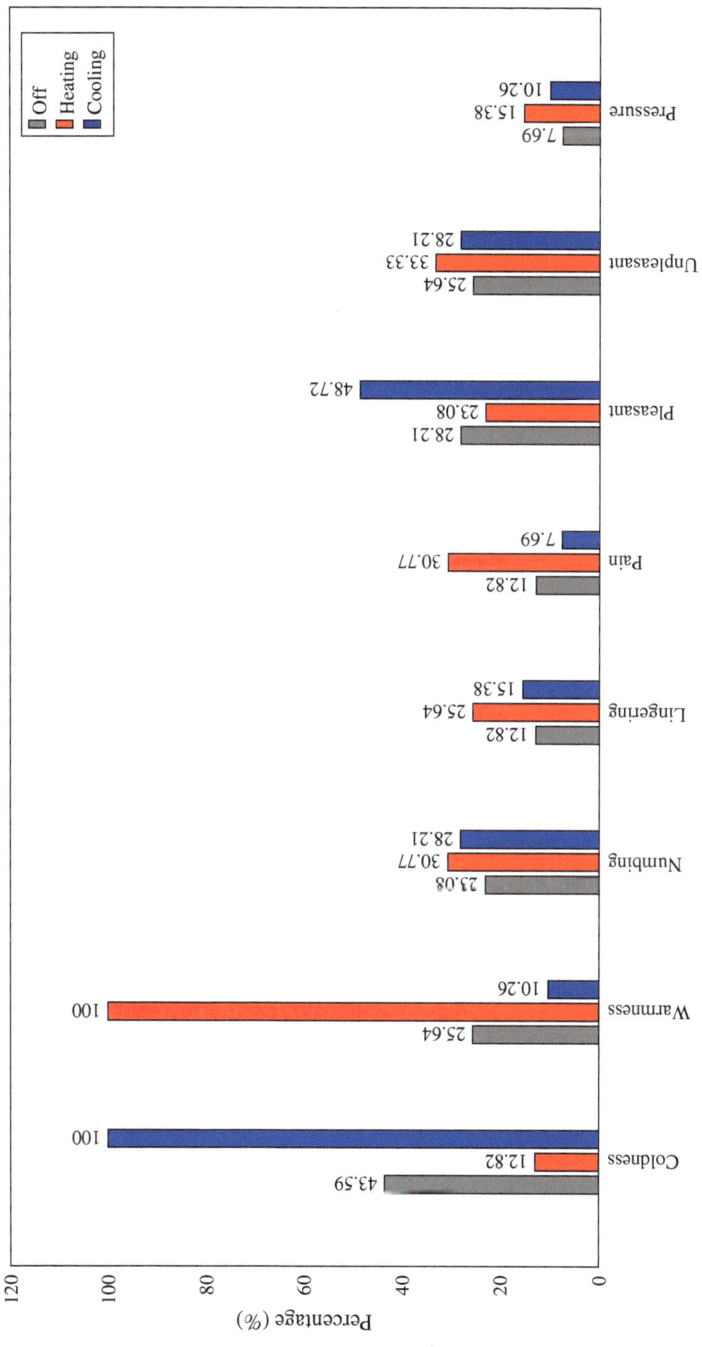

Fig. 5.10 Percentage of participants who reported non-taste related sensations for the three stimulations; off, heating, and cooling

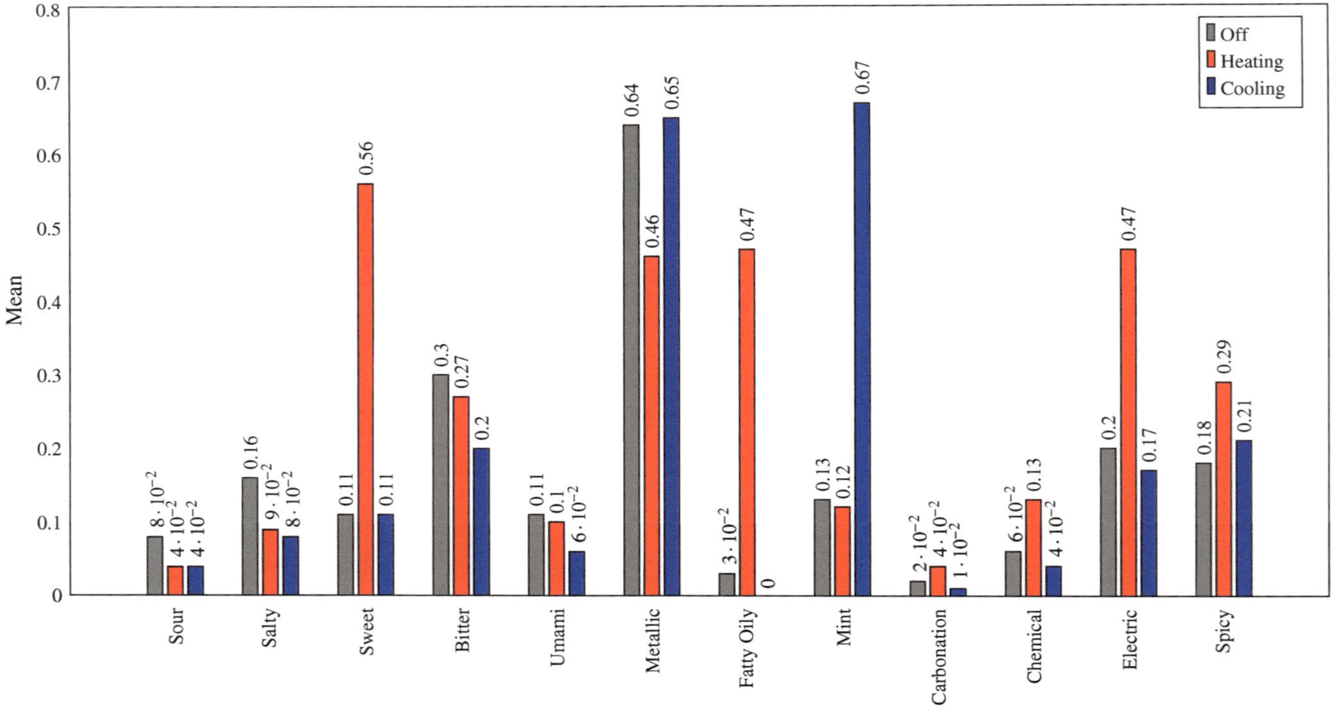

Fig. 5.11 Means of the taste related sensations reported for three different stimulations. We obtained stastically significant results for sweet, minty, fatty, and electric tastes

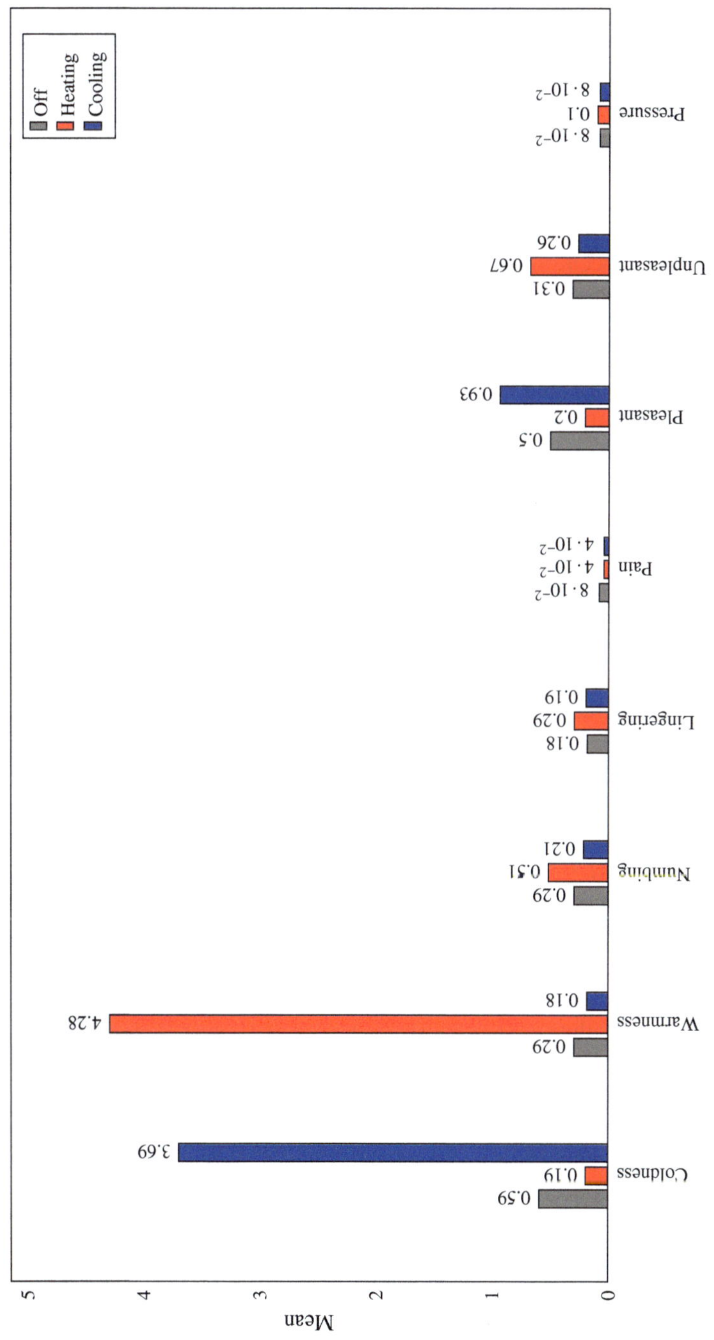

Fig. 5.12 Means of the non-taste related sensations reported for three different stimulations. We obtained stastically significant results for pleasantness, coolness, and warmness

it is stimulated with temperature [20]. Therefore, temperature rise can promote the activation of the channel. Similarly, we hypothesized that stimulating the tongue using thermal taste machine should also increase the sensitivity for sweet taste sensations. Therefore, we decided to conduct an experiment to test this hypothesis using thermal taste machine and sucrose solutions. Twenty volunteers (15 males and 5 females, Mean age = 25.30, SD = 5.43, age range = 20–44) were participated for this study from Universiti Technologi Malaysia. The experiment was conducted in a quiet meeting room at the university.

Three different solutions were used for this experiment including two sucrose solutions (with concentrations of 3.1 gL^{-1}, 24 gL^{-1}) and one mineral water solution. Six different test scenarios were created with the two modes of the device (device on and device off) with the three solutions. The trials were randomized between the participants. In each trial, participants were asked to place the tip of the tongue on the thermal taste interface approximately for 14 s. This was the time taken for heating the tongue from 20 to 40 °C. After the thermal stimulation, participants tasted a solution (randomized between participants) and rated how strong sweet sensation they felt. They also rated the likeness and Intensity of the solution. The range of these scales were from −50 to +50 (+50 = extreme sensation, 0 = neutral, −50 = extremely no sensation). Then participants rinsed the mouth with water to make sure there is no remaining taste solution in the mouth. We repeated these same steps for next 5 trials as well.

However, this experiment was conducted with two limitations. The first limitation was that as soon as the silver plate was removed, the tongue started to cool. The second limitation was that the sweet taste receptors which are located in the other areas, such as the sides and back of the tongue and the soft palate, were not heated. Therefore, the effect of the thermal stimulation was only limited to the sweet taste receptors located in the anterior of the tongue.

Results of the experiment is shown in Fig. 5.13. Thermal stimulation improved the sweetness of the solutions, specially for the 3.1 gL^{-1} sucerose solution, where the means for the device off was rated only as −12.2 and improved up to −2.925 with the device switched on. For the other two solutions, the sweetness rating was also positively improved. Paired T-Tests results for the sweetness for all three solutions regarding the Device On and Off were signeficant ($p = 0.024$ for water, $p = 0.007$ for 3.1 gL^{-1} sucrose solution, and $p < 0.001$ for 24 gL^{-1} sucrose solution). A similar findings was reported previously in the medical field [21]. Similar positive effect was observed for intensity for all three solutions ($p < 0.001$ for water, $p = 0.004$ for 3.1 gL^{-1} sucrose solution, and $p < 0.001$ for 24 gL^{-1} sucrose solution). Participants liked the higher sucrose concentration more and found it sweeter and more intense than the lower sucrose concentration and water. This is the first occurrence that a thermal taste device developed in VR or HCI fields reported enhancement of sweetness. Therefore, our device can be used in VR applications for enhancing the sensation of sweetness in combination with chemical sweeteners. Our results showed that this effect can be applied for both thermal tasters and non-thermal tasters.

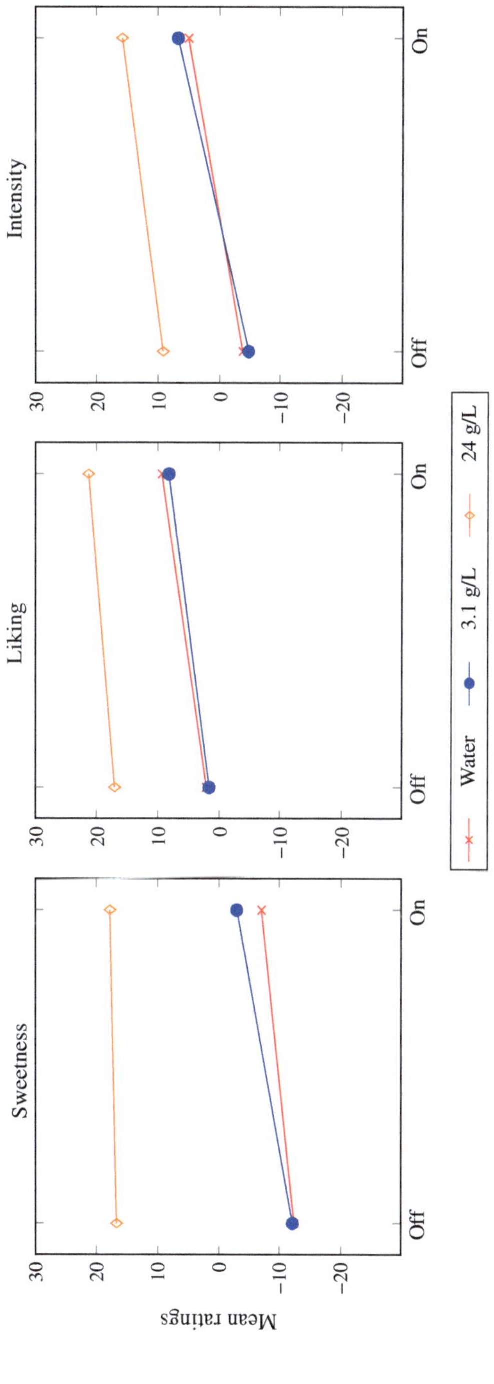

Fig. 5.13 Thermal taste device significantly improved the sweetness, and intensity of the solutions

5.5.3 *User Evaluation of Different Stimulation Speeds Affect the Sweet Sensations*

This experiment was designed as two sub-experiments. During the first part of the study we identified people who could receive thermal sweet sensations from the thermal taste device. Then, in the second part of the experiment, we asked those thermal sweet tasters to try three different rates of temperature rise and rate the sweetness they felt for different rates of temperature change.

5.5.3.1 Experiment to Identify Thermal Sweet Tasters

During the first study, we asked the participants to rate how strong they feel five basic taste sensations (sour, sweet, bitter, salty, and umami) during the thermal taste stimulation. We purposely asked the participants to report their responses for five basic tastes to avoid developing any bias towards sweet taste. The intensities were marked in a scale from -50 to $+50$ (where $+50$ is extreme sensation, $0 =$ neutral, -50 is extremely no sensation). Then we found six participants who reported thermal sweet taste sensations ($M = 13.33$, $SD = 12.11$). Our idea was to study how different stimulation speeds affect sweet taste. Since our participant size for the experiment was only six subjects we treat this experiment as a pretest.

5.5.3.2 Pretest with Different Stimulation Speeds

After identifying six thermal tasters, we stimulated these six participants using the three different stimulation rates slow, medium, and fast (rates were approximately 0.66 $°Cs^{-1}$, 1 $°Cs^{-1}$, and 1.5 $°Cs^{-1}$). Each participant went through 12 trials (3 stimulations $\times$ 4 times). After each stimulation, the participants were asked to remove the interface and evaluate the sensation on three visual analogue scales for intensity,

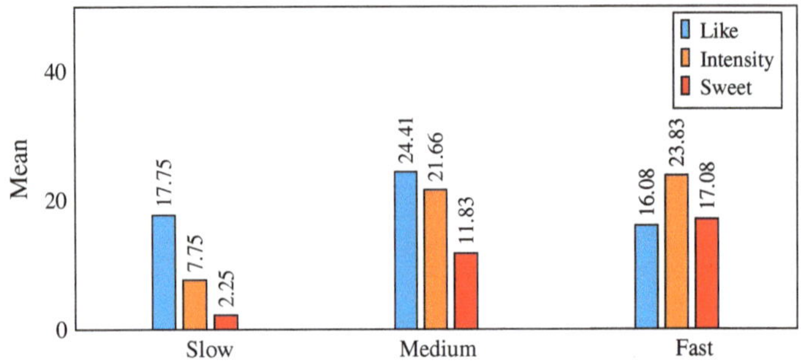

Fig. 5.14 Means of likeness, intensity, and sweetness reported for slow (0.66 $°Cs^{-1}$), medium (1 $°Cs^{-1}$), and fast (1.5 $°Cs^{-1}$) rates of temperature rising

valence, and sweetness (Intensities of the scales were ranged from −50 to +50). Figure 5.14 shows the results of this study. Participants reported more 'sweetness' and 'intensity' (of the sensation) for faster rates of temperature rise. Further, participants 'liked' the medium speed. Since the sample size is smaller for this test we consider this experiment as a pretest. This effect will be further investigated in future.

5.6 Discussion and Future Works

The 'Thermal Taste Machine' showed positive results towards producing and modifying taste sensations for sweet, chemical, minty, fatty, pleasantness, heating, and cooling. Except for the sensations of sweetness, heating, and cooling, this is the first time that any other sensation has been reported using thermal taste stimulation. Therefore, the first experiment successfully showed that thermal taste is a combination of more than one sensation. This study also demonstrated that heating of the tongue produced sweet sensations, which was reported by all the previous works mentioned in the literature review section.

Further, our second experiment proved that sweet taste can be easily enhanced by the thermal stimulation. This is the first time it was found in a thermal taste device developed for VR or HCI. This effect can be used in VR applications to enhance sweetness for both thermal tasters and non-thermal tasters. Finally, our third study suggested that faster temperature rise can produce more intense sweet sensations. This can be directly applied on enhancing the sweetness of virtual food and beverages in VR applications for thermal tasters.

We are currently working on two new prototypes for this device; 1. develop a spoon shape interface that change the temperature of the food we eat and 2. develop a cup like interface where users can adjust the temperature of the drinks. So the basic idea of the spoon interface is to use it when people eats desserts as shown in Fig. 5.15. We placed a tasteless jelly on top of the spoon, user will be asked to heated up the tongue using our device before eating the jelly. We would like to study whether changing temperature of the food can modify the food consumption behavior. We are planning for a long term study where we invite subjects for lunch every week and after the lunch they will be served the dessert with 'Thermal Spoon'. We will have a control group who eats desserts in room temperature (25 °C) and the others may eat the dessert less than the room temperature (10 °C) and higher than the room temperature (40 °C). We already know that according to Talavera's study and referring to our own user study discussed in this paper high temperatures activate more sweet sensations. Also from the other hand people usually eats more quantity of the dessert if the food is not sweet. So our hypothesis would be people will eat less when the desserts are served with high temperature while they will eat more when the desserts are served in low temperature.

The mug interface will be used with drinks where it can maintain a certain temperature. As same as the spoon interface we believe that we could modify the food consumption habits of people who drink something sweet by modifying the sweet-

Fig. 5.15 Desserts demonstration to audience in Brainy Tongue Conference, Spain

ness of the drink. Currently we are experimenting of altering our device to develop these two interfaces. The first study we did was to study how long that the device take up to rise the temperature of the water. With compared to the metal, water has higher specific heat (Specific heat of water is approximately 4200 J/Kg °C while specific heat of copper is 385 J/Kg °C). The results of this preliminary experiment is shown in the Fig. 5.16. Here we used the two different copper plate with the dimensions of 210 × 50 mm and 120 × 50 mm. The time taken to heat up the water is quite longer than 500 s when the device is operated with 40 °C. Therefore, we operated the Peltier with 75 °C (since 80 °C is the maximum temperature it can produce according to the data sheet). Then the copper coil design was able to reached to the 40 °C within 366 s. We can further increase the speed of heating the water by providing more voltage to the Peltier and adjusting the PID control value. We are yet to do this test with changing the copper plate to a silver plate as well.

We believe the thermal taste machine would be very useful in future. In multisensory communication systems, sensory experiences that involve visual, audio, and touch are, broadly speaking, implemented in different contexts successfully. However, sensing and reproducing taste and smell experiences is still a challenge. We believe after conducting further user evaluation tests and improving our device, we would be able to propose proper stimulation parameter set for the reproduction of sweet sensations digitally.

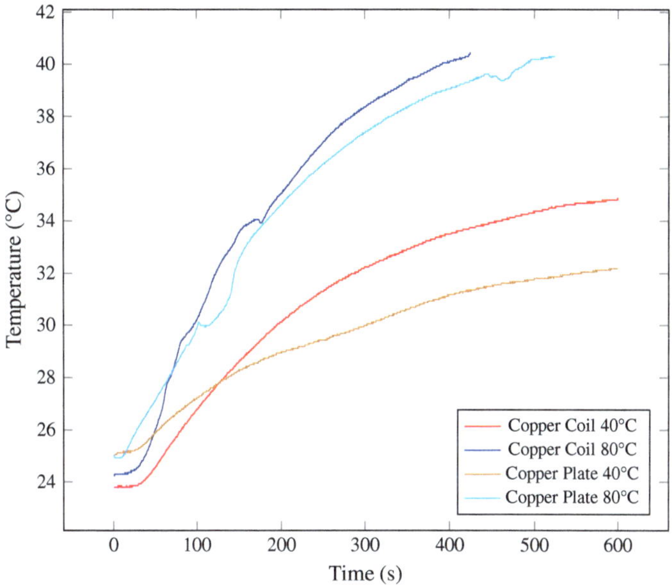

Fig. 5.16 Temperature change rate measurements for the proposed mug prototype

5.7 Conclusion

This research has made the following novel research contributions: 1. The first thermal taste device that can produce sweet, chemical, minty, fatty, and pleasant sensations. 2. We measured 20 types of thermally induced (taste-related and non-taste) sensations and presented how they were modified by heating and cooling. 3. Our device was the first thermal taste device that can enhance the sweet sensations. 4. It has also suggested that faster rates of temperature rise may produce more intense sweet sensations. The final objective of this research is to develop a successful thermal taste technology that can be easily integrated into VR and digital communication. This digital generation of sweet taste will be useful for several fields like gaming, virtual reality, entertainment, and online marketing. We also hope that this research will be useful for certain clinical populations such as, for example, patients with diabetes.

References

1. Beauchamp GK, Cowart BJ (1987) Development of sweet taste. Sweetness 127–140
2. Ranasinghe N, Cheok AD, Nakatsu R (2012) Taste/ip: the sensation of taste for digital communication. In: Proceedings of the 14th ACM international conference on multimodal interaction. ACM, pp 409–416

3. Nakamura H, Miyashita H (2013) Controlling saltiness without salt: evaluation of taste change by applying and releasing cathodal current. In: Proceedings of the 5th international workshop on multimedia for cooking & eating activities. ACM, pp 9–14
4. Suzuki C, Narumi T, Tanikawa T, Hirose M (2014) Affecting tumbler: affecting our flavor perception with thermal feedback. In: Proceedings of the 11th conference on advances in computer entertainment technology. ACM, p 19
5. Kortum P (2008) HCI beyond the GUI: design for haptic, speech, olfactory, and other nontraditional interfaces. Morgan Kaufmann
6. Maynes-Aminzade D (2005) Edible bits: seamless interfaces between people, data and food. In: Conference on human factors in computing systems (CHI 2005)-extended abstracts. pp 2207–2210
7. Tutton M (2008) Designers developing virtual-reality'cocoon'. CNN
8. Cruz A, Green BG (2000) Thermal stimulation of taste. Nature 403(6772):889
9. Talavera K, Yasumatsu K, Voets T, Droogmans G et al (2005) Heat activation of trpm5 underlies thermal sensitivity of sweet taste. Nature 438(7070):1022
10. Green BG, Nachtigal D (2015) Temperature affects human sweet taste via at least two mechanisms. Chem Senses 40(6):391–399
11. Talavera K, Ninomiya Y, Winkel C, Voets T, Nilius B (2007) Influence of temperature on taste perception. Cell Mol Life Sci 64(4):377
12. Ranasinghe N, Karunanayaka K, Cheok AD, Fernando ONN, Nii H, Gopalakrishnakone P (2011) Digital taste and smell communication. In: Proceedings of the 6th international conference on body area networks, ICST (Institute for Computer Sciences, Social-Informatics and Telecommunications Engineering), pp 78–84
13. Ranasinghe RAN (2012) Digitally stimulating the sensation of taste through electrical and thermal stimulation. PhD thesis
14. Cheok AD, Karunanayaka K, Samshir NA, Johari N (2015) Initial basic concept of thermal sweet taste interface. In: Proceedings of the 12th international conference on advances in computer entertainment technology. ACM, p 52
15. Spencer B Specific heat of metals
16. O'Dwyer A (2009) Handbook of PI and PID controller tuning rules. World Scientific
17. Liu P, Shah BP, Croasdell S, Gilbertson TA (2011) Transient receptor potential channel type m5 is essential for fat taste. J Neurosci 31(23):8634–8642
18. Bajec MR, Pickering GJ (2008) Thermal taste, prop responsiveness, and perception of oral sensations. Physiol Behav 95(4):581–590
19. Yang Q, Hollowood T, Hort J (2014) Phenotypic variation in oronasal perception and the relative effects of prop and thermal taster status. Food Qual Prefer 38:83–91
20. Liman E (2007) Trpm5 and taste transduction. In: Transient receptor potential (TRP) Channels. Springer, pp 287–298
21. Yeomans MR, Tepper BJ, Rietzschel J, Prescott J (2007) Human hedonic responses to sweetness: role of taste genetics and anatomy. Physiol Behav 91(2):264–273

Chapter 6
Digital Smell Interface

Abstract Technology in communication is rapidly growth in the past years. Scientist and researchers are competing to give the best way of connecting people through communication. Currently, our daily lives are deeply ingrained with digital communication and the technology now have been develop until all human senses could be digitize to have more interactive experience. Most of our daily activities can be captured, shared and experienced as images, audio, and video using digital devices. When we watch a movie, we experience the stimulation of sights and sounds. Then what about the sense of smell? This project presents the first digital technology developed for transmission of smell through digital networks. The digital stimulation of smell is considered as a useful step in expanding the technology related to multisensory communication. Previous methods for activating the sensation of smell chemically, has obvious disadvantages such as being complex, expensive and lower controllability. Most importantly, smells exist in molecular forms making it impossible to communicate over a distance. Therefore, generating smell sensations without chemicals is becoming highly significant for our increasingly digitized world. We propose a digital interface for actuating smell sensations. This is done by stimulating the olfactory receptors of the nasal concha using weak electrical pulses.

6.1 Introduction

We use so much of communication in our daily lives to transfer information and knowledge. People share and convey the message in the form of communication, verbally or non verbally. The meaning of the communication itself seems to be simple but how the process of the communication to be happen had a various way just to make it possible. Traditional methods of communication had a lot of concrete and specific way such as visually, audibly or need to be both. As time goes by, when places had become the issue to the transferring knowledge process, then there comes the involvement of the technology towards the communication. The communication includes more digital technology and even now the multisensory communication; smell, taste, and touch had been rapidly develop to be inserted to the digital communication. The contribution on multisensory towards communication could give

A. D. Cheok and K. Karunanayaka, *Virtual Taste and Smell Technologies for Multisensory Internet and Virtual Reality*, Human-Computer Interaction Series,
https://doi.org/10.1007/978-3-319-73864-2_6

huge impact to the development of information technology. The communication could be more interesting and realistic even when it happens to be at two or more different places and time through the internet. There are a lot of research had been done to investigate the effect on the contribution of the digitized sensory towards communication but only few focus on transferring the smell digitally.

The early digital smell interfaces developed over the years in the field of Human Computer Interaction are chemical based [1–6]. They used chemicals to generate the odour sensations. These systems are complex, expensive to use frequently, and need routine maintenance for smooth operation. Also there are other limitations such as difficult to switch from one smell to another smell using the device, difficult control the area of stimulation over the time because gas molecules and liquid molecules tend to float around the space and because of the motion of the molecules, it is hard to keep a constant concentration of the smell in a space. Therefore, generating smell sensations without using the chemicals is becoming a need. By achieving that we believe that this research will enable people to experience smell sensations digitally across the internet as we experience visuals and audio in our everyday life.

Digitizing smell is not an easy process since we need to give much attention on how to trigger the smell without the use of any liquid chemicals and gas odorants. It is easy for the liquid and gas to generate a smell where any state of liquid and gas have molecules that can move towards the sensitive receptors inside the nose and trigger the smell. Basically, a smell is generated by triggering the olfactory receptors inside the olfactory epithelium. There are several receptors will be triggered to generate a smell then the smell is transfer through the olfactory nerves to olfactory bulb. The olfactory bulb has a function to communicate with the brain and tells the brain there is a smell there and process the smell.

In order to digitize smell sensation without the using any chemicals, we propose a new user interface that has potential to trigger smells by stimulating the olfactory receptors in the nose non-invasively with weak electrical pulses. In this project we are researching on developing a new interface that can induce weak electrical pulses on the smell receptors and generate smell sensations. The concept of this interface is shown in Fig. 6.1. The main element to generate a smell are depending on these two elements; frequency and current. So we decided to develop a computer controllable user interface which produce weak direct current (DC) electrical pulses in varying these 2 parameters. DC electrical pulses is chose based on its safety towards the use on human. The human body impedance is also said to be higher for DC as compared to AC (cited website). By placing the two silver electrodes of the device, into the nasal concha (as shown in Fig. 6.1) and stimulating the smell receptors, we expect that this interface would produce some type of smell sensations. Area near the nasal concha is ideal to use for the stimulation because of the two factors. The first reason is nasal concha has a high density of olfactory receptors and the second reason is we can easily fix the two electrodes in between the superior and middle or middle and inferior concha.

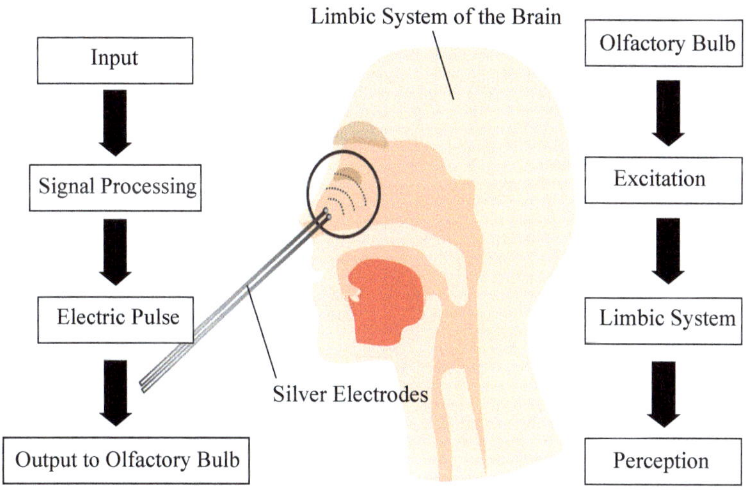

Fig. 6.1 Concept of the stimulation

6.2 Related Works

Till this day, most of the methods used for smell generation is based on the use of odor molecules (chemicals) [2, 3, 5]. There are several other studies on olfactory devices, both with stationary emission [7–9]. However, wearable devices so far uses tubes to send scented air to the user's nose [10, 11]. Some are wearable and movable [12], allowing users to receive personalized scented notifications.

However, these methods have several setbacks such as being complex, invasive, expensive, require maintenance and refilling, the controllability is less and non-distribution in the air. Most essentially, the chemical based smells cannot be transmitted across the digital network and regenerate remotely as it been done for visual and auditory data. But auditory and visual data are based on electromagnetic frequencies and so can be digitized easily unlike smell data which is based on chemical molecules and chemoreceptors. A few examples of olfactory devices using the chemical approach includes Sensorama, olfactometer, Scentee and inScent.

6.2.1 Chemical Based Taste Actuation System

6.2.1.1 Sensorama

Probably the first interactive computer controlled smell system was the Sensorama [13] invented by Morton Heilig in New York, USA. It was introduced in 1962 to provide multi sensory theatre experiences. This machine could stimulate

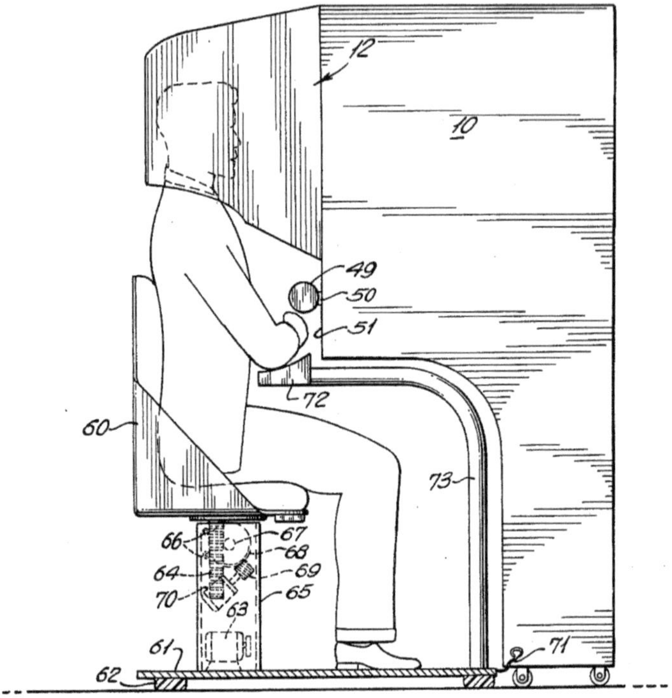

Fig. 6.2 Sensorama simulator [13]

several senses such as visual, hearing, and smell. Furthermore, it has produced movements, vibrations and direct wind effects. Figure 6.2 shows a side view drawing of the Sensorama simulator with the seat, arm rest and controllers. The device provides a housing hood mounted above the seat to fit the head of the user. It has an enclosure scent arrangement where the scent could be directly released into the hood. The scent can be removed and applied in a short time since only the small volume of air is involved.

6.2.1.2 Olfactometers

An olfactometer is an instrument used to detect and measure odor dilution. Human subjects in laboratory settings use olfactometer for research and most times to measure human olfaction. Its major use is to measure the odor threshold of substances. To measure intensity, olfactometers present an odorous gas as a starting point against which other odors are compared. On the other hand, an olfactometer is a device used for producing smells in an accurate and controlled manner [14]. Laboratory olfactometers are specially designed to have different smell channels and a continuous flow channel.

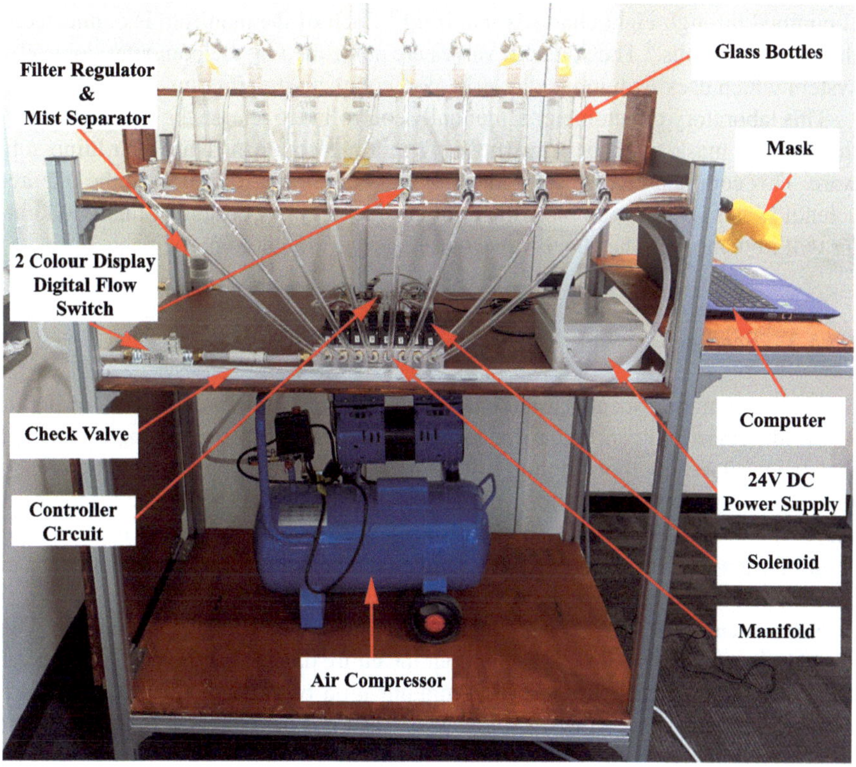

Fig. 6.3 Diagram of laboratory olfactometer

Figure 6.3 shows an example of a laboratory olfactometer built by the authors of this book [15]. The olfactometer consists of three subsystems, the pneumatics system (which contains an air compressor, filter regulator and mist separator, digital flow switches, check valves, solenoids, manifolds, glass bottles, connectors, gas hose clips and Polyurethane tubing), controller circuit and controller program.

In the pneumatic system, an 'Oil-Free Air Compressor'[1] was used to pressurized air in olfactometer. The air compressor can rapidly pump and store air inside the storing tank. The air from the air compressor releases to the olfactometer system with one odor channel with a flow rate of 5 L/min. Then, the air compressor releases the air through air filter, mist separator and pressure regulator unit [16]. These components provide clean filtered air from the dust and water particles. After that, the air flows through the polyurethane tube to digital flow switch.[2] The flow switch is used to control the flow rate by turning the knob at the flow switch. Subsequently, the air

[1]Europower Silent Oil-Free Air Compressor EAX-5030, http://www.mudah.my/Europower+Silent+Oil+Free+Air+Compressor+EAX+5030-36828289.htm.

[2]SMC PFM711-F02-B-W digital flow switch, IFW/PFW FLOW SWITCH, http://www.smcpneumatics.com/PFM711-F02-B-W.htm.

is pumped through eight channels manifold.[3] Each of the manifold is connected to the solenoid valves.[4] The solenoid valves are controlled by the computer controlled system which uses Arduino and a serial port connectivity program.

This laboratory olfactometer is manually controlled to release the specified smells through the mask to participants nose by sending commands from Arduino software. This command will trigger Arduino microcontroller to switch on the required solenoid. This air flow will blow the liquid inside of glass bottles and finally odorant air will flow through the check valve before going straight to participants nose.

6.2.1.3 Scentee

Scentee, is a mobile scent actuation device, plugs into 3.5 mm audio jack of iOS or Android device and can be carried along. It is the world's first phone attachment that can produce smell or fragrance by using smart phone applications. The scent is released using ultrasonic motor from removable reservoir. The device as shown in Fig. 6.4, contained only one aroma at a time and the user must held the device in front of the nose. The scent release is triggered through touchscreen input or via an incoming text message or social network notification.

Scentee can be used as a wake-up alarm, whereby the smell wakes one up. This could influence one's emotion throughout the entire day [17]. With scentee, one can recreate or simulate the emotional communication of an original dish. Therefore, smell has advantages over other senses because it has direct access to emotions and memory.

6.2.1.4 InScent

Another example of olfactory devices is inScent [12], a wearable olfactory display device. This device can be worn in mobile daily conditions. InSent can be worn as a pendant around the neck and can hold about eight different scent aromas that could be inserted and quickly exchanged through small scent cartridges. It allows the user to receive personalized scented notifications.

This device utilizes the properties of smell as a warning channel by amplifying received mobile notifications with artificially discharged scents. As the scent is released, scent aroma will evaporate and it is carried towards the user. InScent has several advantages compared to other olfactory display devices. InScent is wearable and mobile while several other olfactory devices are stationary and not wearable. Also, because the device holds up to 8 different scents user can easily select what scent they need at every given time.

---

[3] SMC VV307-01-083-02-F manifold assy, 8-sta, VT3/VO3 SOL VALVE 3-PORT, http://www.smcpneumatics.com/VV307-01-083-02-F.html.

[4] SMC VV307-01-083-02-F manifold assy, 8-sta, VT3/VO3 SOL VALVE 3-PORT, http://www.smcpneumatics.com/VV307-01-083-02-F.html.

Fig. 6.4 Scentee device which can be programmed to release the smell in the cartridge in a given time (https://scentee.com)

6.2.2 Non-Chemical Based Actuation System

From the above, the possibility of using non-chemical stimulation methods to stimulate smell and taste sensations digitally can be seen. However, the above reported studies are conducted mainly in the medical domain (with controlled environments), are invasive, or only in the experimental stage. Therefore, to achieve electrical stimulation methods as a means of actuating the sensations of smell, this research will achieve research breakthroughs in controllability, accuracy, and robustness.

Electrical activity of the olfactory bulb has been examined for more than a decade. In 1851 Schönbein suggested that people can smell the effects of electric discharge (through ozone) and raised questions about the truly electric smell sensation [18]. Yamamoto has first studied the electrical stimulation of the olfactory mucosa and deep structures of the brain in a rabbit [19]. They were stimulated using 0.2–0.5 ms duration of pulse in the range of 1–20 V in strength. Recording electrodes was placed on the scalp of forehead record a slow negative wave of evoked potential.

Similar research has been performed again with rabbits [20] that used the stimulation parameters of 2 mA of current as 0.5 ms duration of pulse. With the 2 Hz rate of stimulation, the evoked potential showed 2 successive waveforms with peaks. The first experiment to detect olfactory bulb activity using electrical stimulation for human was done by Ishimaru et al. [21]. Electrical stimulation of 2 mA with 0.5 ms, is used and the olfactory bulbar potentials were recorded from the frontal sector of the human head. This experiment has suggested that olfactory nerves are producing the similar waveforms that researchers found in the previous experiments on rabbits.

In a study, Kumar et al. [22] has presented that electrical stimulation of the olfactory bulb and tract produces the pleasant and unpleasant smells in epilepsy patients.

It is the only finding so far that discuss about the type of olfactory hallucinations produce by electrical stimulation. In one of the experiment authors reported that, nine subject perceived unpleasant smells (like bitterness, smoke, or garbage, while two subject perceived a pleasant smell (like strawberry or good food) and lastly, five subject did not smell anything. The stimulation was based on varying the current with three steps; trials starting from 3 mA were increased to 6 and 9 mA. The total stimulation period for a single trial was 5 s with 50 Hz frequency. The pulse duration of a single cycle was 300 μs. The olfactory hallucinations occurred for stimulation of 3 and 6 mA. This is the only experiment that clearly reports electrical stimulation on olfactory receptors generates smell sensations. However, this method cant be used for the general population since it is highly invasive (this method requires electrodes to be implanted in brains frontal lobe).

In 2002, Ishimaru et al. [23] diagnosed the olfactory disturbance to identify olfactory ability in 14 patients including 12 patients who suffers from Parosmia, Anosmia, and Hyposmia. They proposed that electrical stimulation can be used as a substitution for the odor stimulation test such as Toyoda and Takagis perfumists strip method (T&T) olfactometry which has used as a standard for psychophysical olfactometry. Olfactory evoked potential from the scalp was recorded, and the result is analysed between both methods. The electrical stimulation use 2 mA with duration of 0.5 ms and stimulation was at the wider nasal cavity. Some of the evoked potentials of the patients were detectable, some were not due to their health record background. However, there is no sensation of smell experienced by all the test subjects.

To our knowledge, there are still few reports on electrical stimulation of human olfactory mucosa. Early study [24], reported various olfactory sensations resulting from electrical stimulation of the olfactory neuro-epithelium using anodal and cathodal stimulation. The method he claimed resulted in distinct types of smell perception. Induced odor perceptions such as vanilla, almond and bitter almond was produced by anodic stimulation, while cathodic stimulation produced a wide-ranging burnt odor. This is the only study that suggest noninvasive electrical stimulation of olfactory receptors can produce smell sensations. The latest attempt of electrical stimulation of olfactory receptors was made by [25], and this study also failed to produce any smell sensations. Authors have used current ranging from 50 to 800 A with the frequencies of 2, 10, 70, 90, 130 and 180 Hz. Subjects reported visual flashes, pin pricks, and itching sensations near the reference electrode as their experiences. However, subjects reported that there is a significant impact on the perceived intensity of the smell while sniffing an odorant during the stimulation.

6.3 Method

In order to trigger smell from electric stimulation, frequency and current is the important elements that need to be control. The stimulation needs to have few kilohertz of frequency in a certain rate while the output current should be control as not more than 5 mA [26].

6.3.1 Development of Device

The design and development of the prototype describing in this section. The circuit have 2 power supply; one from the Arduino and one from the current controller circuit. Arduino is one of the simple microcontroller for driving electronics circuit. First prototype developed is using 555 timer to produce square wave pulse. As 555 timer is very well known as integrated chip (IC) used for pulse generator but it is not suitable for variable frequency stimulation. Therefore, Arduino is used to generate the square pulse wave in a certain interval of time. Arduino will send a pulsed DC signal as the frequency output and it have several sets of frequency to be choose.

The stimulation current is set by the current controller circuit. A schematic diagram of the current controller circuit is shown in Fig. 6.5. This current controller circuit can generate a constant current to the signal square wave pulses coming from the Arduino. The output current from the Arduino could send as high as 40 mA. To control the current so it has a low steady current, LT3092 programmable current source is used [27]. It requires 2 resistor to set up the output current coming from the Arduino and this component could control the output current ranging from 0.5 to 200 mA.

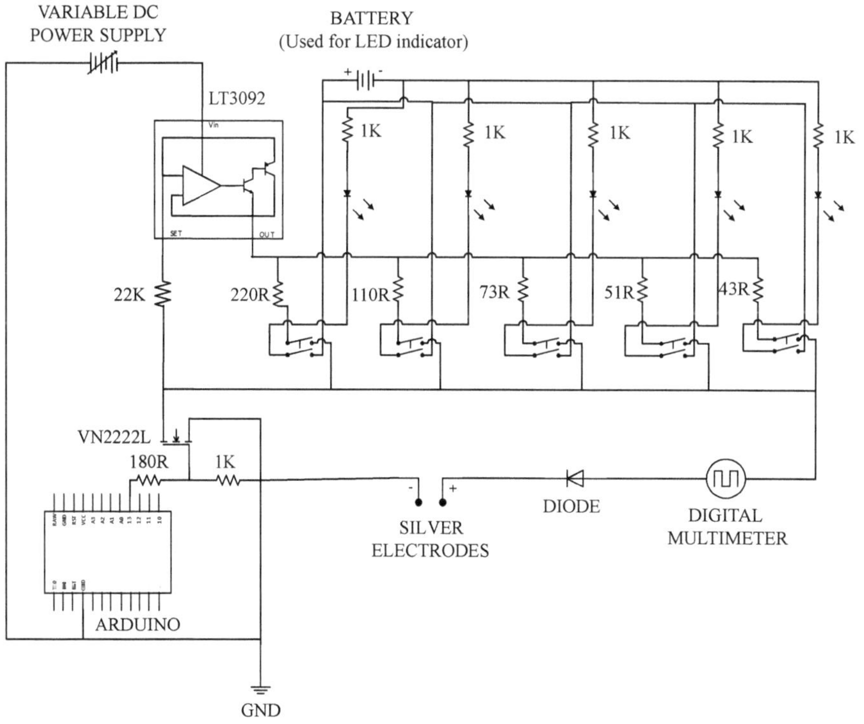

Fig. 6.5 Circuit diagram

In this project, the maximum current need to produce is only from 1 to 5 mA. Here is the calculation used for determine the value of resistors; (R_{set}) and (R_{out}):

$$Iout = \left(\frac{Vset}{Rout}\right) = \left(\frac{10uA.Rset}{Rout}\right) \tag{6.1}$$

consider $Iout = 1, 2, 3, 4, 5\,\text{mA}$.
consider $Rset = 22\,\text{k}\Omega$ (reasonable starting level of voltage across R_{set}, V_{set} is 200 mV).

Therefore, the value of R_{out} is setting up as in Table 6.1.

Fixed current produce by the circuit is driven by the current controller circuit. Current controller mode could regulate a fixed current output according to the calculation desired, despite any voltage given. The amount of current output by the circuit can be controlled using one of the five push buttons shown in Fig. 6.6 and the respective LED near the push button will lights up after the selection.

The maximum load resistance of this operating device is up to 4 kΩ. Power supply produces the variable voltage range from 1 up to 20 V to output constant current flow. Arduino microcontroller accepts the control key sequences received through the USB connection of the PC and drive the current controller board accordingly, based on the experiment protocol. We are currently using a serial monitor program to control this device; however, it is possible to easily integrate this device with a user experiment software that provide serial interface connectivity. The frequency of the stimulation pulses and stimulation time is controlled by the Arduino program. It is possible to vary the stimulation frequency from 0 to 33 kHz.

The stimulation parameters and safety margins for the olfactory receptors was defined by referring the prior works. Our device was purposely limited to control the current between 1 and 5 mA to avoid uncomfortable feelings or damage to the subjects. Current of 2 mA will be the starting current for our first experiment since 1 mA of current is considered as too weak for stimulation, while more than 4 mA could induce some pain and 5 mA of current is considered as the maximum harmless current for the human subjects [28]. The maximum continuous stimulation period (switching on and off interval for the stimulation) is defined as 0.5 s by also referring to the same previous studies [29, 30]. During the stimulation period of 0.5 s, the device will be operated in 2 kHz and stimulate the subject by sending square wave

Table 6.1 Value of R_{out} based on desired current

Current (mA)	R_{out} (Ω)
1	220
2	110
3	73
4	51
5	43

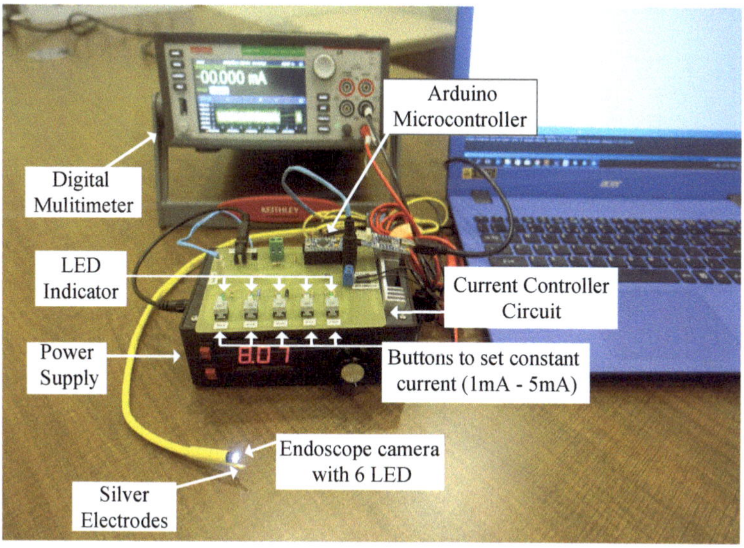

Fig. 6.6 Current controller circuit device

shape pulses to the nasal concha. We can easily modify this frequency by sending an input to the microcontroller firmware using the computer serial interface.

For the initial user studies we have programmed the device to be operated on four frequencies; 2, 10, 70, and 180 Hz. Figure 6.7 shows a sample of different signal patterns generated by the device. For amplitude, the signal has a 2% error due to component tolerance. Besides that, there are negative spikes at the falling edge due to stray inductance. From equation $V = L\frac{di}{dt}$, negative voltage was induced from stray current as there are considerable change in the current value at the transient.

The pair of silver electrodes, Fig. 6.8 which used to stimulate olfactory receptor neurons, are custom made with the dimensions of 100 mm in length, 0.5 mm in width and a sphere tip of 0.8 mm diameter at one end. We kept approximately 1 mm distance between the two electrodes. During the stimulation, one electrode is configured as the positive and the other electrode configured as the ground.

The two sphere tips supposed to make contact with the inner wall of the nose during the stimulation as shown in Fig. 6.9. This device only operates if the electrodes read resistance from the skin. Therefore, during the stimulation, a digital multimeter was to observe the current flow of the device. This was to ensure the electrodes are properly touching the wall of nasal cavity. When two electrodes made the contact with the skin properly, during the stimulation, multimeter outputs the correct amount of current. We found that 5 mA is the maximum harmless current that can be used to stimulate human oflactory receptors [31]. Therefore, we have stimulated the subjects only within 1–5 mA range.

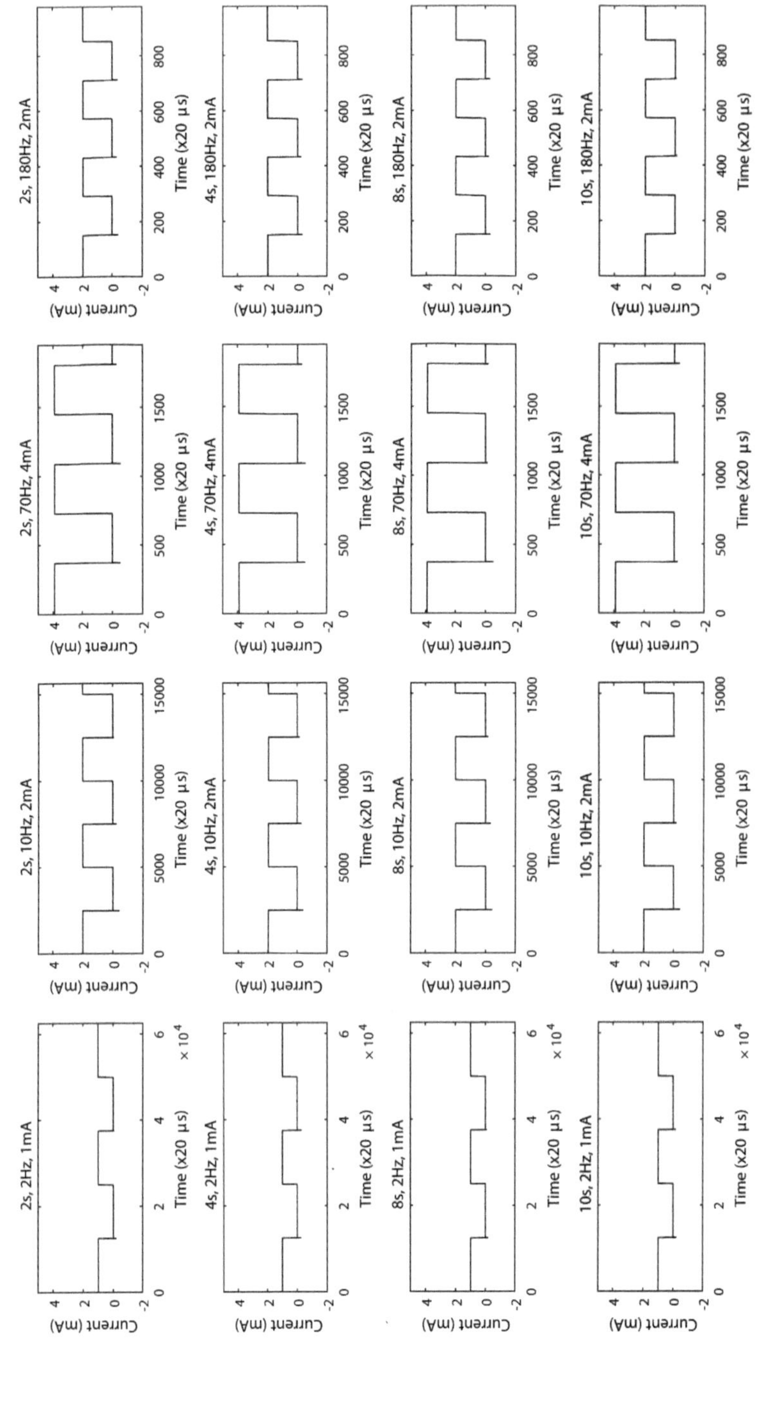

Fig. 6.7 Different stimulation frequency patterns generated by the device

Fig. 6.8 Silver electrodes

6.4 User Evaluation

In this study, diverse sites at the ventral surface of the inferior, middle, and superior turbinate, as well as the septum dorsum were targeted. The different stimulation parameters used in this experiment are shown in Table 6.2. Permutation of those parameters (frequency and current) were used randomly for stimulating subjects. In this study, before investigating the possibility of stimulating smell perception on the subjects, we conducted extensive pilot testing with 14 subjects. The goal of the pilot studies was to familiarize ourselves with proper stimulating electrode placement at various locations in the nasal cavity. First, we practiced only electrode placements, and then included current stimulation. Also, taking into cognition how different people react to both the nasal spray used[5] and the silver electrodes. In the initial testing, it yielded different sensations and few odor perceptions.

6.4.1 *Subjects*

A total of 31 healthy subjects comprising 20 males and 11 females participated in this study. A majority of the participants were university students who were between the ages of 20–23 years old, and a few adults also participated in the experiment (mean age 25 $\pm$ 5.3). Participants filled and signed a consent form before the experiment and the experiment was conducted according to the ethics guidelines approved by the institutional review board (IRB). A questionnaire was first filled by each individual relating to his/her health status and allergies. Participants were physically screened by answering some general well-being questions. The participants also confirmed

[5] Azelastine Nasal Spray, https://www.drugs.com/cdi/azelastine-spray.html.

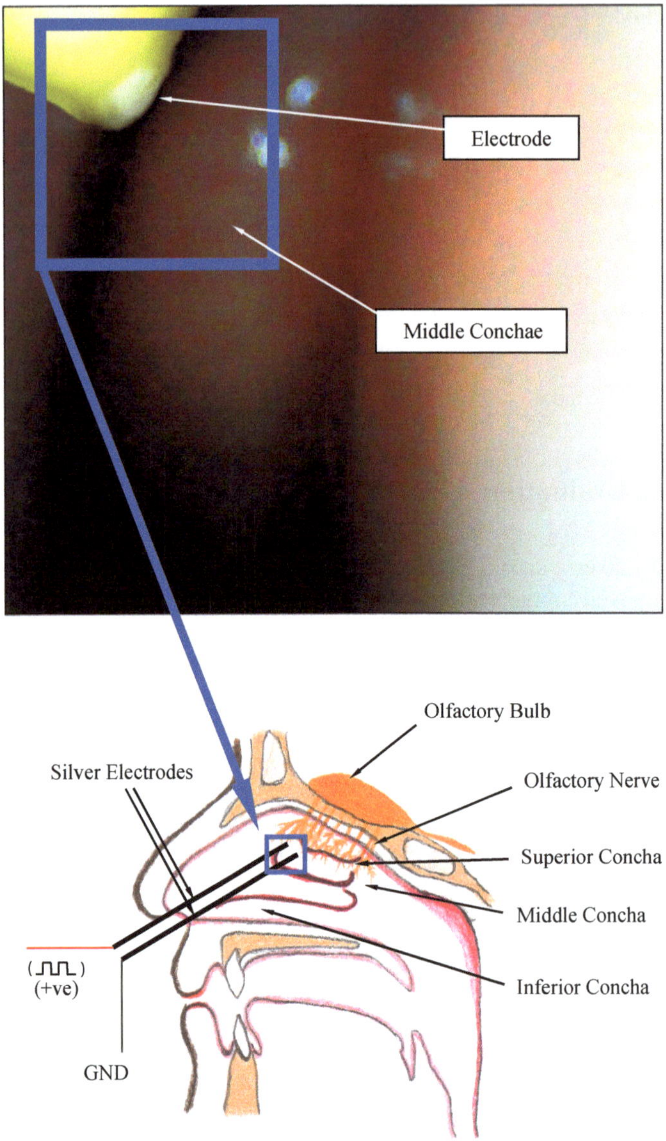

Fig. 6.9 Placement of the electrodes inside nasal concha

whether they have intact olfaction, nasal congestion, non-use of chronic medication of any kind, no recent head injury, and no history of psychiatric disease. After that they participated in the following experiments.

Table 6.2 Different stimulation parameters (currents, frequencies, and time periods) used for the experiment. Permutation of these parameters (as shown in Fig. 6.7) were used randomly for stimulating subjects

Current (mA)	Frequency (Hz)	Time (s)
1	2	2
2	10	4
3	70	6
4	180	8
5	180	10

6.4.2 *Pre-screening Experiment: Sniffing of Known Odorants Aimed at Ascertaining the Smelling Capability and to Study the Effect of Electrical Stimulation on the Nasal Cavity*

After completing the questionnaire, subjects were randomly presented with five different known odorants. These odorants were presented in a non-translucent container to conceal the color and numbered 1–5. Participants were asked to sniff each odorant one at a time by closing either the left or right nose, identify the type of odorant, and rate the intensity using a graded scale. After each trial, participants were asked to sniff water. The idea behind this step was to reset the olfactory receptors to their original state. Then, this same set of procedures was repeated for all five odorants. This was done to ascertain the capability of the subject to identify different smells, and to rule out anasomia before proceeding to the main experiment. Finally, after the main electrical stimulation experiment, participants were made to sniff the same odorant again.

6.4.3 *Main Experiment: Electrical Stimulation Directed at Inducing Odor Perception*

Here, we aimed to electrically stimulate the nasal mucosa to evoke odor or other perceptions. Participants were made to seat comfortably on a chair in a closed room. To minimize sneezing and to decongest the nasal cavity, Azelastine [32] nasal spray was used. An eye mask was used to cover the eye of the participants to minimize distractions. They were asked to breath normally, as in natural sniffing., then the stimulating silver electrodes were gently inserted into the nasal cavity with endoscope camera guidance and gradually brought into contact with the targeted areas, as shown in Fig. 6.10. The insertion was preceded by electrical stimulations done concurrently with nasal inhalation as though natural sniffing [33, 34].

Stimulation duration was fixed at 10 s, with an inter-stimulus interval of 60 s. Nevertheless, each stimulation parameter started with an initial current of 1 mA and was continually increased until 5 mA. The participant was free to stop the experiment

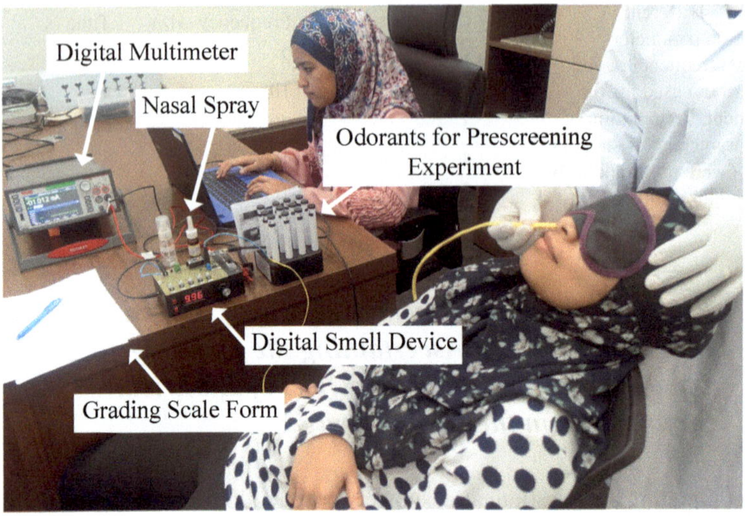

Fig. 6.10 Subject

if they felt the stimulation was uncomfortable. After each trial, subjects were asked to report any types of sensations and rate the intensity using a graded scale ranging from 0 (none) to 10 (strong). The procedure for each participant lasted for about 45 min.

6.5 Results

6.5.1 The Response Generated by Electrical Stimulation of the Nasal Mucosa

Each participant underwent eight trials. During the interval between trials, participants completed a response form where they indicated the type of sensation experienced during and after each trial and rated the intensity using a graded scale. The participants completed the response form for the eight trials, after which an analysis for each of the participants responses were made, and the parameters compared. The responses were divided into two categories: smell sensations and olfactory-induced sensations. The first category consists of the 10 basic smells, as described by [35], and the second category consists of the 12 induced non-olfactory sensations. Sensations such as tingling, pinpricks, numbness, pain, pressure, lingering, heat, burning, metallic, electric and cooling were recorded.

We started with the lowest current and frequency before increasing both gradually. Parameters of 1 mA at 10 Hz and 1 mA at 70 Hz were the stimulation parameters which gave the most prominent results for the smell related responses. Figures 6.11 and 6.12 represent these smell perceptions based on [35] categorization. Electrical stimulation with 1 mA and 70 Hz induced the highest odor perception with an average of 0.74, +1.6 on the 10-point grading scale. 27% of the participants reported that they perceived fragrant and chemical sensations. Other reported smell sensations include: 20% fruity, 20% sweet, 17% toasted and nutty, 10% minty, and 13% woody. The non-olfactory-induced sensations for 1 mA and 70 Hz (illustrated in Fig. 6.13) include: 67% pain, 53% tingling, 50% heating, 50% electric, and 50% pinprick.

Other stimulation parameters such as 1 mA and 10 Hz (shown in Fig. 6.11), a mean of 0.65 $\pm$ 1.4. Here, 17% reported fragrant, 27% sweet, 10% chemical, and 10% woody. Results for the 4 mA and 70 Hz stimulation is shown in Fig. 6.14 where the mean sensation for pain is 2.95 $\pm$ 0.85; in percentages 82% reported pain, while 64% reported pressure. One way ANOVA was conducted to compare effects of electrical stimulating parameters to generate smell sensations and other induced sensations. The variations of pressure and tingling across different parameters are shown in Figs. 6.15 and 6.16.

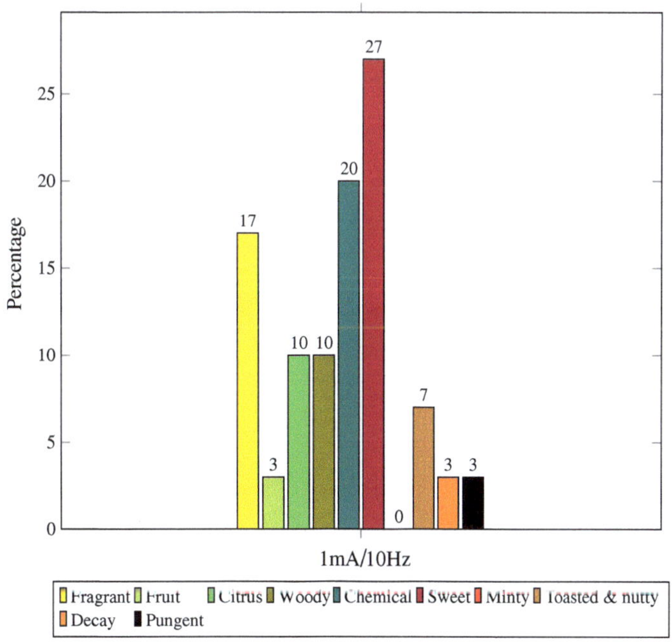

Fig. 6.11 Smell related sensations reported for electrical stimulation at 1 mA and 10 Hz as percentages

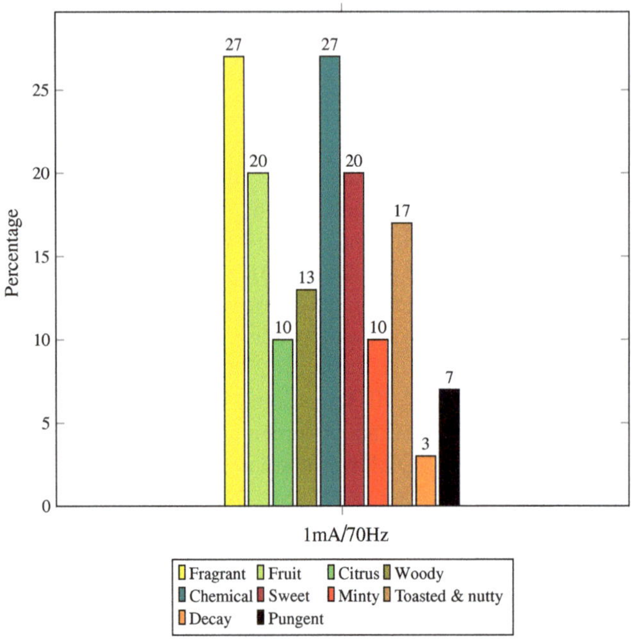

Fig. 6.12 Smell related sensations reported for electrical stimulation by participants at 1 mA and 70 Hz as percentages

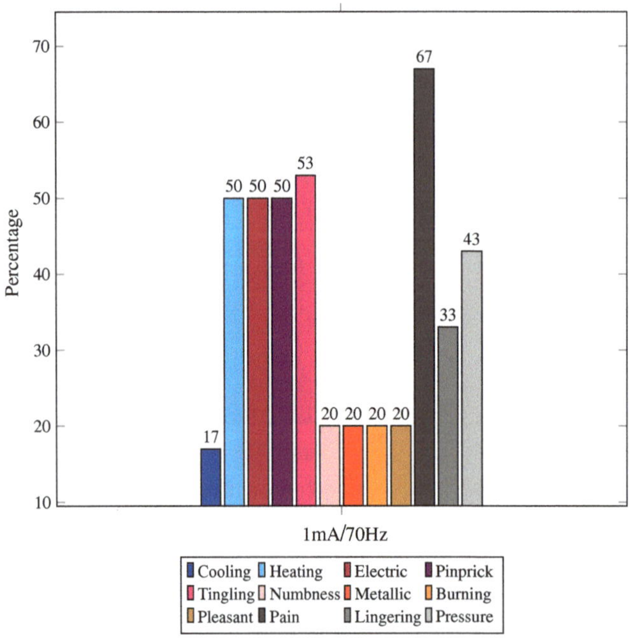

Fig. 6.13 Other sensations reported for 1 mA/70 Hz

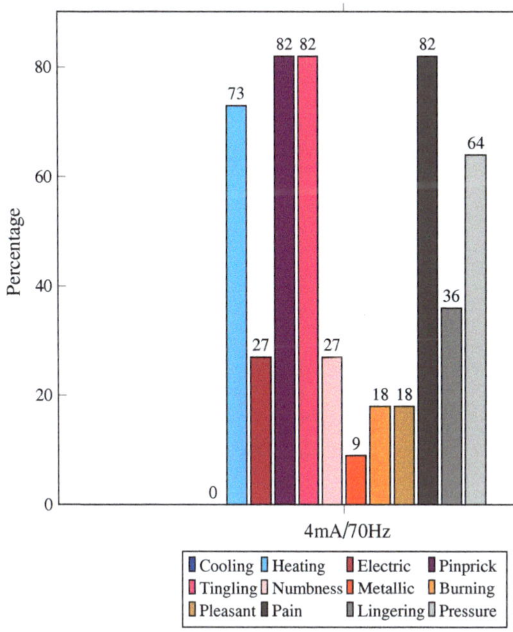

Fig. 6.14 Other sensations reported for 4 mA/70 Hz

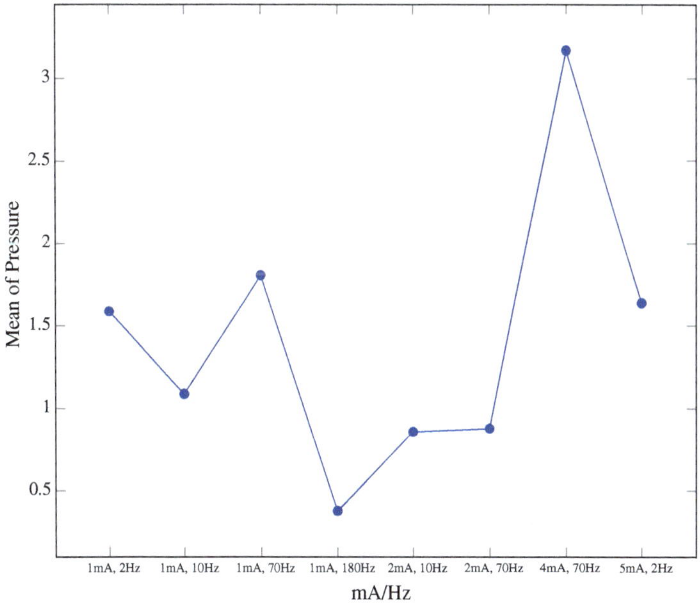

Fig. 6.15 Mean intensities of the pressure induced by different type of stimulations calculated using a range 0–10 (where 10 is the most)

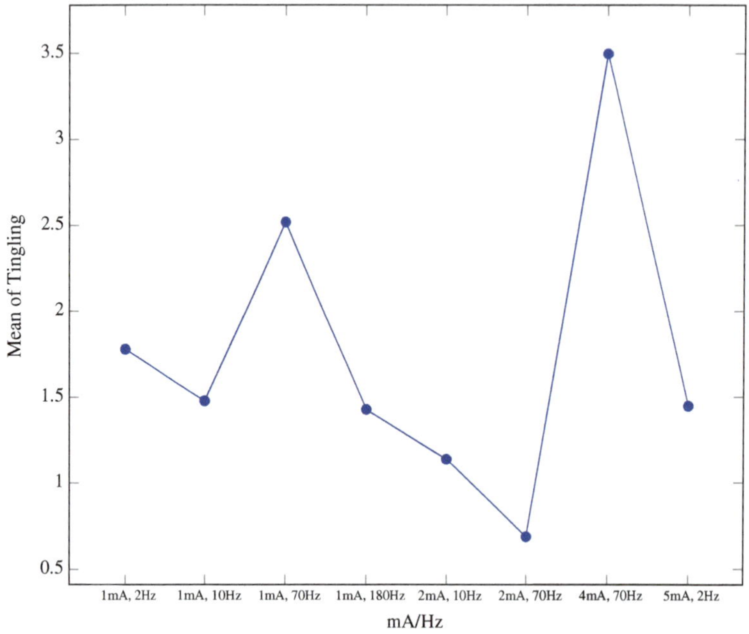

Fig. 6.16 Mean intensities of the tingling induced by different type of stimulations calculated using a range 0–10 (where 10 is the most)

6.5.2 The Response Produced After Sniffing of Known Odorants

We compared the intensity of various odorants before and after electrical stimulation to identify the effect of this stimulation on the nasal mucosa. The capability of each participant to recognize various kinds of odorants was tested. The five known odorants presented to the participants were orange, cinnamon, banana, pineapple, and peppermint. Figure 6.17 compared the odorant intensity before and after electrical stimulation of the nasal turbinates. As show in the graph, the intensity before and after stimulation is similar for all odorants, although most subjects verbally reported they felt a difference.

An independent sample t-test was conducted to determine if there were changes in intensity between before and after electrical stimulation. There was no significant difference in odorant intensity in before and after stimulations. The mean and standard deviation for the odorants are shown in Fig. 6.17. For orange $t(60) = 0.74$, $p = 0.47$, for banana $t(60) = 0.93$, $p = 0.354$, for pineapple $t(60) = 0.62$, $p = 0.95$, for peppermint $t(60) = 0.60$, $p = 0.55$ and for cinnamon $t(59) = 1.18$, $p = 0.24$.

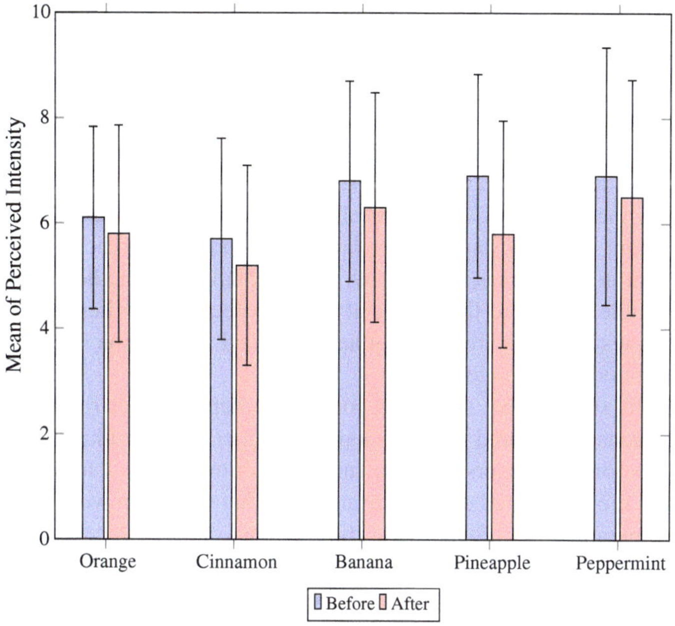

Fig. 6.17 Variation of the intensity of smell for five known odorants before and after electrical stimulation. Error bars shows standard deviations

$P > 0.05$ for all the odorants. The results show that electrical stimulation of the nasal cavity did not have any statistically significant effect on the intensity of the odorants.

6.6 Discussion

Our stimulations were majorly at the middle turbinate where olfactory local field potentials are easily acquired [36], while few were at the superior and inferior turbinates, based on the results from epithelial biopsies on spread of olfactory mucosa [37, 38]. Stimulation at 1 mA and 70 Hz generated induced sensations such as tingling and pressure. However, stimulations at 1 mA and 70 Hz and 1 mA and 10 Hz combinations produced smell sensations such as fragrant, chemical, and sweet smells. Some participants reported perception of smells. This indicates that there might be evidence that there may be an electrical path to stimulate the sense of smell in humans.

Pain was also induced in participants, with stimulation of 1 mA 180 Hz inducing the least pain sensation, while stimulation of 4 mA and 70 Hz induced the highest pain sensations. Few participants reported visual flashes at stimulation of 4 mA and 70 Hz. A previous study [25] also reported visual flashes at 10 Hz and 0.45 mA. Furthermore, we think combinations of high frequencies and high currents induces more pain. At the same time, we observed that pain also occurred when the stimulating electrodes were outside the olfactory cleft and touching the respiratory mucosa.

Furthermore, as a next step, we are planning to study the anodical stimulation and cathodical stimulation in nasal cavity [39, 40]. Specially with anodic stimulation [25], we will be able to reduce the pain that subjects feel. An earlier experiment [25] suggested that sensations are different and electrical stimulation had a mild impact on concurrent odor perception.

The results suggest that there is a connection between electric stimulation and smell sensations. Compared with previous researchers, we are likely to discover the ideal stimulation parameters. For the future experiments, we can use participants who already reported smell sensations and with that subgroup we will try different stimulation parameters. If the electric stimulation would produce smell sensations, we will be able to communicate smell experiences through the internet and regenerate smell sensations digitally. Also, this method will be used as an alternative for sensory restoration [41] for the people who have lost the sense of smell due to some medical conditions.

We are planning to extend this user experiment with more participants. The effects of the different electrical stimulation parameters, such as frequency, current, and stimulation period will be more closely studied in future. By analyzing the results, we plan to identify various stimulation patterns that can produce different smell sensations. If we can show conclusive evidence for effectively reproducing smell sensations, the next step would be comparing the difference between electrical smell perceptions and natural smell perceptions. This can be done by studying the areas of the brain being activated by electrical stimulation and chemical stimulation. If both stimulation techniques activate approximately same areas of the brain, we will be able to argue that electrical stimulation can reproduce the same sensation as chemical based smells do. Also, we are planning to do cell compatibility testing on the electrodes used for electrical stimulation of the human olfactory epithelium. The main objective of this research is to study long term effects of electric stimulation on smell receptor cells.

6.7 Conclusion

Smell is one of the most important sense of the human. Therefore, finding a digital means to activate the sense of smell is becoming an important need specially for future digital communication. In this paper we have presented our proposed technology of using electrical stimulation to excite olfactory receptors using a non-invasive method. During the pretests, subset of participants have reported that they sensed different sensations of smell and differences in perceived intensity for smell. We believe in future, we can improve this technology and clarify exact parameters which are responsible for producing the smell sensations. Our vision is to transform multisensory communication into a new era of Internet with digitizing the sensc of smell as other human senses (sight, hearing, touch and taste) and allow people to create, communicate, and regenerate true multisensory information over the internet.

References

1. Bodnar A, Corbett R, Nekrasovski D (2004) Aroma: ambient awareness through olfaction in a messaging application. In: Proceedings of the 6th international conference on Multimodal interfaces. ACM, pp 183–190
2. Choi Y, Cheok AD, Roman X, Sugimoto K, Halupka V et al (2011) Sound perfume: designing a wearable sound and fragrance media for face-to-face interpersonal interaction. In: Proceedings of the 8th international conference on advances in computer entertainment technology. ACM, p 4
3. McGookin D, Escobar D (2016) Hajukone: developing an open source olfactory device. In: Proceedings of the 2016 CHI conference extended abstracts on human factors in computing systems. ACM, pp 1721–1728
4. Ramos G, Boulos M, Balakrishnan R (2004) Pressure widgets. In: Proceedings of the SIGCHI conference on Human factors in computing systems. ACM, pp 487–494
5. Seah SA, Martinez Plasencia D, Bennett PD, Karnik A, Otrocol VS, Knibbe J, Cockburn A, Subramanian S (2014) Sensabubble: a chrono-sensory mid-air display of sight and smell. In: Proceedings of the 32nd annual ACM conference on human factors in computing systems. ACM, pp 2863–2872
6. Warnock D, McGee-Lennon M, Brewster S (2011) The role of modality in notification performance. In: IFIP conference on human-computer interaction. Springer, pp 572–588
7. Washburn DA, Jones LM (2004) Could olfactory displays improve data visualization? Comput Sci Eng 6(6):80–83
8. Yanagida Y, Noma H, Tetsutani N, Tomono A (2003) An unencumbering, localized olfactory display. In: CHI'03 extended abstracts on human factors in computing systems. ACM, pp 988–989
9. Nakaizumi F, Noma H, Hosaka K, Yanagida Y (2006) Spotscents: a novel method of natural scent delivery using multiple scent projectors. In: Virtual reality conference, 2006. IEEE, pp 207–214
10. Narumi T, Nishizaka S, Kajinami T, Tanikawa T, Hirose M (2011) Augmented reality flavors: gustatory display based on edible marker and cross-modal interaction. In: Proceedings of the SIGCHI conference on human factors in computing systems. ACM, pp 93–102
11. Yamada T, Yokoyama S, Tanikawa T, Hirota K, Hirose M (2006) Wearable olfactory display: using odor in outdoor environment. In: Virtual reality conference, 2006. IEEE, pp 199–206

12. Dobbelstein D, Herrdum S, Rukzio E (2017) Inscent: a wearable olfactory display as an amplification for mobile notifications. In: Proceedings of the 2017 ACM international symposium on wearable computers. ACM
13. Heilig LM (1962) Sensorama simulator (August 28 1962) US Patent 3,050,870
14. Beavers J, McGovern T, Adler V (1982) Diaprepes abbreviatus 1: laboratory and field behavioral and attractancy studies 2. Environ Entomol 11(2):436–439
15. Karunanayaka K, Saadiah H, Shahroom H, Cheok AD (2017) Methods to develop a low cost olfactometer for multisensory, psychology and neuroscience experiments. In: Proceedings of the 43rd annual conference of the IEEE industrial electronics society, Beijing, China. ACM
16. SMC (2017) Smc ac30d-02cg-a fr/ms combo modular, ac mass pro
17. Schredl M, Atanasova D, Hörmann K, Maurer JT, Hummel T, Stuck BA (2009) Information processing during sleep: the effect of olfactory stimuli on dream content and dream emotions. J Sleep Res 18(3):285–290
18. Schönbein CF (1851) On some secondary physiological effects produced by atmospheric electricity. Medico-chirurgical Trans 1:205–220
19. Yamamoto C (1961) Olfactory bulb potentials to electrical stimulation of the olfactory mucosa. Jpn J Physiol 11(5):545–554
20. Ishimaru T, Sakumoto M, Kimura Y, Furukawa M (1996) Olfactory evoked potentials produced by electrical stimulation of the olfactory mucosa. Auris Nasus Larynx 23(1):98–104
21. Ishimaru T, Shimada T, Sakumoto M, Miwa T, Kimura Y, Furukawa M (1997) Olfactory evoked potential produced by electrical stimulation of the human olfactory mucosa. Chem Senses 22(1):77–81
22. Kumar G, Juhász C, Sood S, Asano E (2012) Olfactory hallucinations elicited by electrical stimulation via subdural electrodes: effects of direct stimulation of olfactory bulb and tract. Epilepsy Behav 24(2):264–268
23. Ishimaru T, Miwa T, Shimada T, Furukawa M (2002) Electrically stimulated olfactory evoked potential in olfactory disturbance. Ann Otol Rhinol Laryngol 111(6):518–522
24. Uziel A (1972) Stimulation of human olfactory neuro-epithelium by long-term continuous electrical currents. J Physiol 66(4):409–422
25. Weiss T, Shushan S, Ravia A, Hahamy A, Secundo L, Weissbrod A, Ben-Yakov A, Holtzman Y, Cohen-Atsmoni S, Roth Y et al (2016) From nose to brain: Un-sensed electrical currents applied in the nose alter activity in deep brain structures. Cereb Cortex 26(11):4180–4191
26. Mayerhoff E (2016) "The electric shock questions" #bingo #chi2005. BLOG (2005). http://www.highvoltageconnection.com/articles/ElectricShockQuestions.htm. Accessed Mar 2016
27. Linear Technology (2009) 200mA 2-terminal programmable current source. Rev C 2
28. Fausto N, Campbell JS, Riehle KJ (2012) Liver regeneration. J Hepatol 57(3):692–694
29. Scully SM (2014) The animals that taste only saltiness
30. Roper SD (2013) Taste buds as peripheral chemosensory processors. In: Seminars in cell and developmental biology, vol 24. Elsevier, pp 71–79
31. Mayerhoff E (2005) The electric shock questions, effects and symptoms
32. Astelin A (2017) Azelastine nasal spray
33. Rojas-Líbano D, Kay LM (2012) Interplay between sniffing and odorant sorptive properties in the rat. J Neurosci 32(44):15577–15589
34. Mainland J, Sobel N (2006) The sniff is part of the olfactory percept. Chem Senses 31(2):181–196
35. Castro JB, Ramanathan A, Chennubhotla CS (2013) Categorical dimensions of human odor descriptor space revealed by non-negative matrix factorization. PloS one 8(9):e73289
36. Lapid H, Hummel T (2012) Recording odor-evoked response potentials at the human olfactory epithelium. Chem Senses, bjs073
37. Rawson NE, Ozdener MH (2013) Primary culture of the human olfactory neuroepithelium. In: Epithelial cell culture protocols, 2nd Edn, pp 81–93
38. Féron F, Perry C, McGrath JJ, Mackay-Sim A (1998) New techniques for biopsy and culture of human olfactory epithelial neurons. Arch Otolaryngol-head Neck Surg 124(8):861–866

39. Uziel A (1973) Stimulation of human olfactory neuro-epithelium by long-term continuous electrical currents. J Physiol 66(4):409–422
40. Vogel R (2016) Understanding anodal and cathodal stimulation
41. Fleiner F, Lau L, Göktas Ö (2012) Active olfactory training for the treatment of smelling disorders. Ear Nose Throat J 91(5):198

Chapter 7
Discussion and Conclusion

7.1 Discussion

Emerging from the multisensory subject, our goal is to transform Internet communication into a multisensory experience by digitization of the smell and taste senses and as other human senses (sight, hearing, and touch) and allow people to create, communicate, and regenerate this multisensory information over the Internet. This chapter will discuss the general advantages, future use and some limitations associated with digital taste and smell. To start with, advantages of this technology are presented for various fields including entertainment, virtual reality, medical, gaming, and communication. The technology of sharing smell and taste via Internet is a challenging experience. Presently, there are few technologies that uses chemical to generate smell [1, 2] and taste [3, 4] sensation, but these technologies are complex, needs routine maintenance for easy operation and are expensive to use regularly and therefore the need to digitally stimulate smell and taste.

7.2 General Advantages of Digital Taste and Smell

Smell and taste sensations are very vital part of human living. Furthermore, personal experiences can be related with these sensations. Individuals want to eat together and organize different food items based on their preference for occasions and celebrations in their regular daily activities and will as well want to have a memory of the events. Therefore, the smell and taste sensations are particularly important to sustain a healthy life.

A. D. Cheok and K. Karunanayaka, *Virtual Taste and Smell Technologies for Multisensory Internet and Virtual Reality*, Human-Computer Interaction Series,
https://doi.org/10.1007/978-3-319-73864-2_7

Currently, daily interaction with digital media through the Internet plays an important role in our lives. Humans can express and utilize their complete senses when they interact face to face. Nevertheless, right now, digital communication need to depend on a constrained range: sound, text and image only or in combination of two or three [5]. Our studies as described in this book has explored the integration of smell and taste into digital medial without the use of chemical to improve its interaction.

7.2.1 Multisensory Digital Communication

In digital communication media, sense of smell and taste are remarkably given minimal consideration. In the field of multisensory communication, digitization of smell and taste sensations are of great benefit. Currently, digital communication are only based on text, audio, and video. Therefore, advancing the effectiveness of these technologies as well as faster connectivity has been the focus of several research works. Yet, these technologies presently available does not increase the intimacy amongst the parties involved in the communication. As a result, in this current paradigm of communication, multisensory communication has become essential to enhance the qualities of life [6]. Apart from that, it will enlighten the benefits of moving forward into the current era of information involving non-verbal expressions such as smell, taste, and not only auditory or visual. Smell and taste will change the way of communication in the future. Posting a photo on social media with the sense of smell and share our experiences together with friends and family will become possible. As an example, sharing a taste of cake with family and friends over the Internet during a birthday celebration will provide a memorable experience. In a family where members are geographically not together, sharing their everyday events is important to build a stronger family relationship.

7.2.2 Online Shopping

Today, many people prefer to shop online owing to limited time and energy. Online shopping is primarily based on visual information only (the design, size and color of the product). What about adding an extra sensory cue to attract the attention of customers to make a choice, such as smell, especially buying perfumes and some food products. If each of the perfumes transmits the specified odor to the users during shopping, it will be able to help and eases the process of deciding what item to purchase. When there is a new fragrance produced, rather than seeing only images, smells can be sent to notify the customers that they have received a new scent. Besides that, a combination of smell and taste during online shopping will increase the percentage of Internet users buying online food. During this time, just by looking at food images, users can only guess and bring back a flood of memories about the taste of a food. Seeing, smelling and tasting the food product will triggered user to make appropriate choices because they can now compare between several options.

7.2.3 *Virtual Reality*

Presently, Virtual Reality (VR) systems involve the stimulation of two or three senses, sight, hearing and touch. VR experience can be more enjoyable by adding other senses such as olfactory and gustatory. The main goals of VR are to create a realistic perception and virtual environment that can be felt in real-time. Unfortunately, the VR cannot simulate the reality perfectly and have some limitations. An entire new set of experience could be presented by utilizing the digital taste and smell technology. Currently, immersive VR is deliberated as a future innovation, in the same way the technology of digitizing smell and taste could be utilized as the immersive to produce modified mental state by intensely connecting ones sense of smell and taste.

7.2.4 *Entertainment*

The technology of smell and taste sensation can be used for example in taste or smell based communication entertainment, by allowing controllability of smell or taste digitally. Currently, because of several reasons the entertainment aspects of smell and taste sensation is not investigated. Some of the reasons are absence of controllability of the senses, need of a variety of chemicals, and rapid flexibility for sensations.

However, in the future, utilizing the digital smell and taste technology, a new entertainment idea may be considered. Think of the possibility of being able to smell or taste the cheesy slice of pizza being eaten by your preferred character on TV? Imagine watching a celebrity chef cooking on TV and being able to smell or taste it in the comfort of your home. We believe addition of smell and taste to the television will not only increase but will improve the users' entertainment experience also.

7.2.5 *Medical*

Another benefit of our technology is in the medical field. We hope that this technology would be used for taste and smell disable patient. Malfunctions related to smell such as anosmia, hyposmia, parosmia and phantosmia and taste related malfunctions such as dyseusia, hypergeusia, hypogeusia, heterogeusia, norgeusia, phantogeusia, cacogeusia, parageusia and ageusia [7]. Though these situation are not generally common, nevertheless studying these malfunctions with digital smell and taste technology respectively could permit a different sensory pathway for patients. Digital smell technology might help a patients with Parkinson's disease who loss their sense of smell to be able to smell different odors again. Likewise, digital taste technology might as well assist when a patient is instructed not to eat certain food ingredients example sweet for diabetics and salty for patients suffering from congestive heart failure.

7.3 Limitations

We have noticed several limitations both in stimulation of electric and thermal taste, and in electric smell studies which we deemed as appropriate to mention here. These limitations still needs to be addressed so as to make the electrical and thermal device ready for use. The major limitation being that the device is not user friendliness and acceptability. Because of concern for hygiene and fear of burning the tongue, some users are reluctant to put the silver plate on the tongue. Hence, finding a new way of delivering the electrical or thermal stimulations to the tongue becomes a necessity. Another limitation is the slow onset of thermal taste relative to chemical taste. Waiting time is usually about 10–20 s before taste sensation is felt. Generally, thermal taste sensations are rated weaker compared to taste sensations produced by chemicals. Thus in this research, future contribution would be to find methods that can produce quick and more strong thermal taste sensations. We believe most likely by testing with several temperature ranges, we can improve the rate of temperature change, which consequently improve the intensity of the thermal taste producing different sensations. Presently our stimulation range is restricted for 10 °C up to 40 °C. Experts from different discipline such as flavors and food, nutrition, and medicine would benefit from the improvement of this technology. At that point we would have the capacity to deliberately choose the most essential kind of taste sensations that would profit for individuals and deliver them utilizing this technology.

Similarly there are limitations encountered with electric smell stimulations. In electrical stimulation of smell, the main limitation encountered is the difficulty in inserting the electrode. Because the electric smell stimulation device is not user friendly and some users are hesitant to allow for the insertion of the electrode. The 5 mm diameter endoscope camera attached together with the electrode was considered big by some users and could cause pain. Therefore, maintaining contact between the electrode and the olfactory mucosa is a challenge. And because we tried several parameters on each participant, which we think this also caused pain sensation to participants. Another limitation is the difficulty to identify smell. Because there are no basic smell, it is difficult for users to specify what type of smell stimulated. This was not encountered in electric taste stimulation because there are basic taste qualities (sweet, sour, salt, bitter and umami). In a future study, to avoid pain, smaller endoscope camera for guidance will be used; amount of current will be reduced and series of parameter combinations will be employed.

7.4 Future Plan

Our aim is to develop these technologies as miniature, portable and wearable units. Therefore, enabling users to wear these interfaces in daily lifestyle situations for augmentation. Digitizing taste and smell will provide opportunities for number of new applications. For example, a friend could send you a taste or smell over the Internet by means of a social network and you could taste or smell it electronically. Furthermore, these devices will be able to provide stimulation of a taste and smell from something that cannot exist. This could also lead to a breakthrough in molecular gastronomy. People can make new kinds of recipes using purely digital compositions. We also believe patients who cannot consume certain chemicals such as those with diabetes can stimulate the certain chemical in question (like sugar) taste in place of the actual chemicals. This may allow humans to even consume certain foods that were previously unappetizing to eat or to encourage children to eat unpopular foods to maintain a healthy lifestyle.

7.5 Conclusion

With the goal of better understanding of science of olfaction and taste, we have described several aspect of the two senses. Furthermore, throughout this book we have discussed number of possible technologies including digital smell interface, electrical and thermal taste interface we used which can bring the change from chemical based smell and taste to digital. The technologies, their applications and benefits were also discussed. Furthermore, this book serves as a reference that combines the concepts, descriptions, experimental methods for digitizing the sense of smell and taste. We hope these blue sky quantum step technologies will further improve and will be used in future of Internet communication and virtual reality.

References

1. Choi Y, Cheok AD, Roman X, Sugimoto K, Halupka V et al (2011) Sound perfume: designing a wearable sound and fragrance media for face-to-face interpersonal interaction. In: Proceedings of the 8th international conference on advances in computer entertainment technology. ACM, p 4
2. Seah SA, Martinez Plasencia D, Bennett PD, Karnik A, Otrocol VS, Knibbe J, Cockburn A, Subramanian S (2014) Sensabubble: a chrono-sensory mid-air display of sight and smell. In: Proceedings of the 32nd annual ACM conference on human factors in computing systems. ACM, pp 2863–2872
3. Kortum P (2008) HCI beyond the GUI: design for haptic, speech, olfactory, and other nontraditional interfaces. Morgan Kaufmann
4. Maynes-Aminzade D (2005) Edible bits: seamless interfaces between people, data and food. In: Conference on human factors in computing systems (CHI'05)-extended abstracts, pp 2207–2210

5. Hertenstein MJ (2002) Touch: its communicative functions in infancy. Hum Dev 45(2):70–94
6. Freeman J (2009) The tyranny of e-mail: the four-thousand-year journey to your inbox. Simon and Schuster
7. Henkin R (2004) Taste and smell disorders, human. In: Encyclopedia of neuroscience. Elsevier, Amsterdam, pp 2010–2013

Epilogue

We live in an age when new ideas and inventions in science and technology appear at an astounding rate. As these novelties are converted into products and services they enhance our lives in various ways, some of them practical, some of them pleasurable, some benefiting our health. Behind every such novelty is an individual or a team who discovered or invented the novelty. These geniuses are the creators of our future and should be suitably recognized as such. But sometimes the early stages of their ideas, their work and their research projects might seem strange to the wider public, who are not able to grasp at first the potential of the novelty.

Now and then a star of outstanding technical originality appears in the world of invention, and Adrian Cheok is one such star. Adrian has a passion for devising new ways, both to harness and to enhance the human senses, enabling our senses to detect novel experiences. In this book Adrian and his team at the Imagineering Institute in Johor, Malaysia, describe their work as the pioneers of taste and smell technologies that have been artificially created by digital and electronic means, as well as their work in the development of digital devices for the detection of taste and smell. Using digital technologies this team is striving to enable us to transmit smell via the Internet and to create taste electronically. Imagine the future of cinema, to take just one example, when the audience can smell and even taste the food which an on-screen character is enjoying, a future when a chef in one country can have their latest creation instantly critiqued by a panel of their peers and other cookery experts located throughout the world.

Adrian has pioneered these technologies in his roles as Professor of Pervasive Computing at City, University of London, and Director of the Imagineering Institute. His initial research in this field spawned more media publicity for City, University of London than any other research in its history. And following on from his digital taste and smell discoveries in London, Adrian has founded and led a research team in Malaysia to take his ideas forward.

In this book we find a comprehensive coverage of the field as it is in 2018, as well as an in-depth survey of developments during the few short years when Adrian's ideas were first conceived and expanded. The book will not only provide indispensable background reading for students and researchers of the artificial senses, but it will

A. D. Cheok and K. Karunanayaka, *Virtual Taste and Smell Technologies for Multisensory Internet and Virtual Reality*, Human-Computer Interaction Series,
https://doi.org/10.1007/978-3-319-73864-2

also act as inspiration for those who are looking for an exciting new research domain to investigate. Whatever the future holds for artificial taste and smell, this book is the beginning of that future.

London, 26th January 2018 Dr. David Levy

Biography of Dr. David Levy

David Levy studied Pure Mathematics, Statistics, and Physics at St. Andrews University, Scotland, from where he graduated with a B.Sc. degree. He taught practical classes in computer programming at the Computer Science Department of Glasgow University, before moving into the world of business and professional chess playing and writing. (He wrote more than thirty books on chess). He was selected to play for Scotland in six World Student Team Chess Championships (1965–1970) and in six Chess Olympiads (1968–1978). In 1968 and 1975 he won the Scottish Chess Championship. He was awarded the International Master title by FIDE, the World Chess Federation, in 1969, and the International Arbiter title in 1976.

The development of David's interest in Artificial Intelligence started with computer chess, which was a logical combination of his addiction to chess and his work in the field of computing. In 1968 he started a bet with four Artificial Intelligence professors, including John McCarthy who in 1955 had coined the phrase Artificial Intelligence, that he would not lose a chess match against a computer program within ten years. He won that bet, and another one for a further five years, succumbing only twenty-one years after making the first bet, and then to a forerunner of the program that defeated Garry Kasparov in 1997. David was first elected President of the International Computer Chess Association (ICCA) in 1986, and after a gap from 1992 to

1999 was elected once again, a position he has held since then (the association now being named the International Computer Games Association (ICGA).

Since 1977 David has led the development of more than 100 chess playing and other microprocessor-based programs for consumer electronic products. He still works in this field, leading a small team of developers based mainly in the UK. David's interest in Artificial Intelligence expanded beyond computer games into other areas of AI, including human-computer conversation. In 1994 he brought together a team to investigate pragmatic solutions to the problem, resulting in his winning the Loebner Prize competition in New York in 1997. He won the prize again in 2009.

Davids achievements in the associated field of Social Robotics include founding international conferences on the subject, and being a co-organizer of six such conferences between 2007 and 2017. He has published a primer on A.I., *Robots Unlimited.* His fiftieth book, *Love and Sex with Robots*, was published in November 2007, shortly after he was awarded a Ph.D by the University of Maastricht for his thesis entitled Intimate Relationships with Artificial Partners.

David has had a lifelong interest in organising mind sports events, and was one of the organisers of the World Chess Championship matches in London (1986 and 1993), as well as the World Checkers Championship match between the human champion and a computer program (1992 in London and 1994 in Boston), in addition to dozens of computer chess championships and similar events. In 1989 he inaugurated the Computer Olympiad, for competitions between computer programs playing thinking games, which has since become an annual event. David also created the Mind Sports Olympiad, in which human players compete at more than 30 different strategy games and other "mind sports".

His hobbies include classical music, and he has recently started playing chess again after a long gap away from active play. He lives in London with his wife and their cat.

Index

A. D. Cheok and K. Karunanayaka, *Virtual Taste and Smell Technologies for Multisensory Internet and Virtual Reality*, Human-Computer Interaction Series,
https://doi.org/10.1007/978-3-319-73864-2

MIX
Papier aus verantwortungsvollen Quellen
Paper from responsible sources
FSC® C105338

If you have any concerns about our products,
you can contact us on
ProductSafety@springernature.com

In case Publisher is established outside the EU,
the EU authorized representative is:
Springer Nature Customer Service Center GmbH
Europaplatz 3, 69115 Heidelberg, Germany

Printed by Libri Plureos GmbH
in Hamburg, Germany